D0908758

# Running Today's Factory

## A Proven Strategy for Lean Manufacturing

By
Charles Standard and Dale Davis

Society of Manufacturing Engineers
Dearborn, Michigan

Standard, Charles, 1961—
    Running today's factory: a proven strategy for lean manufacturing
    / by Charles Standard and Dale Davis.
        p.    cm.
    Includes bibliographical references and index.
    ISBN 0-87263-513-9
    1. Factory managment. 2. Production management. I. Davis, Dale, 1945—
    II. Title.
    TS155.S7575  1999
    658.5--dc21                                                       99-33973
                                                                          CIP

Society of Manufacturing Engineers
Customer Service
One SME Drive, PO Box 930
Dearborn, MI 48121-0930
www.sme.org
800-733-4763

1 2 3 4 5 6            01  00  99

# Dedication

To our dear parents
    Mary Alice and Tom
    Mary Elizabeth and Hal

# Table of Contents

# Acknowledgements

This book evolved from our own factory experience, our research in engineering and anthropology, and a comprehensive perusal of the published literature from the fields of engineering, manufacturing, business, operations research, organizational theory, and cultural studies. Countless individuals, scholars, and factory personnel through their articles, books, seminars, conversations, collaborations, and interactions have shaped the perceptions, theories, and insights we present. We extend heartfelt appreciation to all who have helped us clarify our thinking about the science of *Running Today's Factory*.

We acknowledge the early and continued work of Richard Schonberger; it has been a solid foundation for our own work. We are grateful to John Shook whose wisdom and perceptiveness have profoundly shaped our own thinking about lean manufacturing and its role in today's factory. We thank Jeff Liker, a teacher, adviser, and colleague, whose ability to synthesize experiences and data from many sources gives him a comprehensive understanding of the issues and on-going debates in manufacturing management. We thank Rachel Zhang whose research in operations management has allowed us to apply new and provably optimal solutions to complex factory situations. Her thoughtful review of several early chapters contributed to the accuracy of this book. Other key researchers who have helped us understand the underlying determinants of factory behavior include Mark Spearman, Wallace Hopp, and Izak Duenyas.

We sincerely appreciate our many associates in the manufacturing community who have accepted and applied the strategies and principles presented in this book, even when the rationale was theoretical, abstract, or simply based on our own confidence and enthusiasm. Specifically we thank Bill Adams, Dave Olsen, Roger Sprunger, and Tom Tuttle for sharing their valuable insights and allowing us to benefit from their extensive business and manufacturing experience.

We also acknowledge Kelly Callendar, Rick Dixon, and Mike Collins, respected lean practitioners, who help us remember that, in spite of the inherent difficulties, running today's factory can be an enjoyable and even delightful experience. Thanks to Mark Nelms whose pragmatic approach to production management (whether making hydraulic lifters in an automated production facility or American Lager in a five-gallon jug) is a breath of fresh air.

*Designer Clipart 12,000*™ (© 1996 The Learning Company Inc. and its licensors) was the source of illustrations. We are deeply grateful to Laurel Masten Cantor, a dear friend, for applying her creativity and artistry to the design of this book's cover.

We thank Peggy Jennings, our colleague at the American Supplier Institute, who has advised us on many matters and supported in every way our consulting and training endeavors. We express special appreciation to Tom and Mary Standard who kindly and painstakingly edited early versions of our manuscript and, along with Mary Davis, offered encouragement throughout the project. Finally, we express our deep gratitude to Woody Chapman of Hanser Gardner Publications for having the vision, confidence, and patience to guide us through this project and take *Running Today's Factory* from idea to reality.

# Foreword

The Toyota Production System has ushered in a new paradigm in manufacturing now being called "lean manufacturing" in the Western world. A paradigm is a way of thinking. It colors what we see and how we see it. It is a system of beliefs about how the world works. The Toyota Production System was developed at a time when Toyota needed to do more with less, on flexible lines, and reduce the time from order to delivery in order to get paid quickly. The emphasis was on reducing the time line by eliminating anything that did not add value to the product—eliminating *muda* or waste. The most serious form of *muda* in the TPS paradigm is overproduction—producing more than is needed and storing it as piles of inventory. Thus, a huge stamping press that keeps busy by building large batches of product to achieve economies of scale looks very efficient in the mass production paradigm, but absolutely horrible under lean manufacturing. The same machine operating in the same way is looked at completely differently.

Managers in a wide variety of industries throughout the world are finding that lean manufacturing is a better way than mass production. Doing seemingly simple things like taking machines grouped by process and creating one-piece flow cells that make product families can have dramatic effects on work-in-process, productivity, flexibility, and quality. Tying a whole factory together and making just what the customers want when they want it can generate an enormous competitive advantage. Few firms have created a lean enterprise that ties lean factories together with lean logistics, but the results can be phenomenal.

With the large number of successes and the broad acceptance of lean manufacturing as a better way, one would think everyone would be doing it. But quite the contrary, one must search far and wide to find good examples of lean factories. It is easy to find specific elements of lean such as mistake-proofing devices, good workplace organization, work teams, and total productive maintenance programs. Everybody has at least one cell, and some plants even have kanban systems between a couple of operations. But finding an entire lean system with all the elements working in concert still remains a rare treat.

Why are we still in the mass production dark ages in most of today's factories? There are many constraints that make the lean journey a challenging one. At least part of the challenge is understanding the basic lean concepts. *Running Today's Factory* can be of great help in increasing your understanding of lean manufacturing. While there are many books in print that teach in great detail the tools of lean, and a few others that give case examples of the transformation process, it is difficult to find a good explanation of the core concepts of lean manufacturing.

*Running Today's Factory* toes a fine line between providing a theoretical treatment of lean concepts and providing intuitive case examples. And it does this very successfully! It goes just deep enough for the intelligent reader to understand why the philosophy of lean is so powerful. It gives a historical perspective so we can understand how and why this new paradigm came into existence. The authors summarize some very complex concepts of "factory physics" in simple terms that helps us understand scientifically why lean manufacturing works. It is full of facts and figures culled from a broad range of publications, and it uses clever analogies ranging from golf to cooking to help us intuitively understand the concepts. The chapters on pull systems and mistake proofing are filled with practical advice on how to make these difficult tools work in practice. And the authors bring their own experiences in factories implementing lean into the narrative to bring the system to life.

If you want to go beyond blindly copying the tools of others, and move toward understanding the principles of lean manufacturing, this book is for you. It is not a cookbook, but then again anyone who really understands lean would be suspicious of anything claiming to be a cookbook. *Running Today's Factory* will help you understand lean as a system of production and how the various tools of lean contribute to the operation of that system. With this understanding, you can go forth and intelligently implement lean concepts in a way that is appropriate for your factory.

Jeffrey Liker
University of Michigan

*Chapter 1*

# Introduction

*Running Today's Factory* focuses on proven strategies for manufacturing management. We emphasize underlying principles so that specific practices can be understood holistically and applied confidently. Lean manufacturing is presented as a systematic philosophy of production, and we demonstrate how lean principles can be "discovered" independently through rigorous scientific analysis. Using this *science of manufacturing* approach, we illustrate how lean thinking leads to good manufacturing decisions that can be backed up with sound scientific reasoning.

The manufacturing strategies presented in this book are appropriate for any size company including small and mid-size factories. This is of critical importance since more than 96% of all U.S. manufacturers have fewer than 250 employees (U.S. Bureau of the Census, 1996). The increasing customer demand for quality, innovation, flexibility, and low cost presents greater difficulties for small companies primarily due to resource constraints. Therefore, it is particularly important for small manufacturers to increase their productivity and competitiveness (Rishel and Burns 1997). Adopting the strategies presented in this book is one way for small companies to gain a tremendous advantage. In fact, our own experience and current research indicate that small manufacturing concerns are realizing much greater benefits than their larger competitors (Inman and Mehra 1990: 37).

We chose the title of this book, *Running Today's Factory*, quite deliberately. In the late 1970s, the American manufacturing community began to appreciate the manufacturing capability of Japanese competitors; and engineers, managers, and researchers tried to discover the *secret* of Japanese manufacturing. In a landmark article in the *Harvard Business Review* entitled "Why Japanese Factories Work," Robert Hayes stated, "The modern Japanese factory is not, as many Americans

1

believe, a prototype of the factory of the future . . . *it is the factory of today running as it should"* (1981).

## Applicability

The strategies presented in this book are applicable in a remarkably wide range of industrial and nonindustrial situations. Consider for a moment the challenge of producing citrus plants for the citrus growers of the world. The time required to produce a tree has traditionally been between 24 and 36 months depending on specific conditions. By manufacturing standards, this is an incredibly long production lead time! As we will see, the citrus nursery industry faces many of the same challenges that today's factory faces, and this long production lead time makes these challenges even more difficult to overcome.

Consumer tastes (in the literal sense of the word) change constantly as new citrus products become available. Some products come into favor, or flavor, while others become less popular. For example, the sales of pasteurized juice surpassed concentrate juice for the first time in 1998. The changing popularity of tangerines, oranges, and grapefruit in all their different varieties makes it extremely complicated for the grove owner to know what to plant. It is even more difficult for the nursery owner to know what kind of seeds to plant 24 to 36 months in advance. Furthermore, the quantity of seedlings needed in any given year is even more unpredictable. If a citrus region is hit by a freeze, demand can increase by an order of magnitude quite literally overnight! Trying to predict how many plants will be needed two or three years in advance is completely impossible.

The challenge to the nursery owner is to produce the right amount of the right trees at the right time to satisfy the needs of the customers (the grove owners). This is not unlike the challenge faced by today's factory. The nursery owner's challenge is also complicated by extreme variability in customer demand, a diverse and growing product mix, and a long and unpredictable production lead-time. Again, these are no different from the challenges today's factory managers face.

More than 20 years ago, before the American automobile industry had heard about "just-in-time" and before the term "lean production" had been coined, these manufacturing challenges were overcome by one man in the citrus industry of Florida. William Adams, President and CEO of Adams Citrus Nursery in Haines City, Florida, realized that these "challenges" would become almost inconsequential if the *time* required to produce a tree could be reduced to only a few months.

He developed "continuous flow" greenhouses in which seedlings would move from one end to the other through an environmentally controlled area on a special rail system. He invented the Citripot®, which eliminated replanting of seedlings and greatly reduced the material handling effort. The Citripot® was designed to fit onto special "cars" that rolled on the rail system. Seedlings were literally put on the cars at one end of these greenhouses to emerge from the other end as finished trees several months later. From the end of the rail line, they rolled onto a special train that transported the finished trees from the greenhouse to the shipping dock. At the dock the cars were rolled into an insulated "sprinkler van," which transported them to the customer. After the seedlings were placed in the greenhouse, they were untouched by human hands until the customer planted them in the grove.

This method of citrus tree production reduced the time required to produce a tree to nine months—a 75% reduction in lead-time! Over 1.5 million trees could be processed per year using this approach. Adams Citrus Nursery quickly became the world's largest citrus tree producer supplying customers throughout the Americas and as far away as Saudi Arabia. The production methods pioneered by Adams provided him with the flexibility to respond quickly to changing customer demand. Production costs were reduced dramatically because the time and effort traditionally associated with raising citrus trees for transplant were greatly reduced.

It is interesting that the production strategies implemented by Adams were completely consistent with and *almost indistinguishable from those espoused by lean philosophy*. Yet, they were developed and implemented before the first English language article was ever written about Toyota's just-in-time system. We have found that the principles and strategies for running today's factory are almost universally applicable, even though the specific practices or techniques may change depending on the circumstances.

## Lean Techniques

There is a strong cultural tendency for Americans to look for a single and immediate "best solution" (Hammond and Morrison 1996: 169), and lean manufacturing is certainly not immune. Phrases like "zero inventory," "quick setup," "visual factory," "mistake-proofing," and "eliminate waste" are catchy aphorisms. They are also tempting solutions to chronic production problems. However, the results are inconsistent when these techniques are applied as independent, quick fixes. Many manufacturers are experimenting with popular lean techniques such as 5-S, work cells, *kanban* systems, and JIT deliveries. Some are realizing great benefits, while others see no measurable improvement at all. This disparity raises questions about the validity of lean techniques. Are the results predictable? Are the benefits guaranteed?

In fact, lean techniques are not a panacea. Extensive scientific research, industry evidence, and our own shop floor experience indicate that lean techniques alone do not confer any measurable improvement. Lean manufacturing is not a collection of best practices from which manufacturers can pick and choose. It is a production philosophy, a way of conceptualizing the manufacturing process from raw material to finished goods and from design concept to customer satisfaction. Lean is truly a different way of *thinking* about manufacturing.

Many noted authorities, such as Shingo, Deming, and Ohno, caution against imitating best practices or instituting copycat production systems (Pilkington 1998: 31). There is even strong empirical evidence that imitating another company's successful lean system can actually be *less effective* than standard policies (Ocana and Zemel 1996: 212).

So, if lean practices alone do not confer a competitive advantage and if imitation doesn't work, what is the benefit of lean manufacturing? *The tremendous benefits of lean manufacturing are realized when a lean strategy is used to support well-conceived business goals.*

## Manufacturing Strategy

A manufacturing strategy is much more than a slogan such as "eliminate waste" or "reduce setup time." A manufacturing strategy is a pervasive theme or *mission* that guides decision making and establishes direction. Manufacturing strategy is characterized by a consistent pattern of decision making regarding such issues as organization of the factory, equipment acquisition, production goals and measures, and what capabilities the factory develops or eliminates. When the manufacturing strategy is well aligned with corporate strategy, the company realizes a strong, competitive advantage. More importantly, when manufacturing capability helps guide the business strategy, the factory becomes a powerful, competitive weapon. Together, manufacturing strategy and manufacturing capability can provide profound and sustainable competitive advantages that are not easily duplicated.

## Cycle Time and Variability

Pragmatically speaking, the objective of lean manufacturing is to streamline the flow of production material throughout the value stream. This is accomplished primarily by reducing both *cycle time* and the *variability of cycle time* wherever possible. Cycle time is the length of time production material spends in the factory. *Processing time* is the length of time required to process an item at a particular workstation. Sometimes "cycle time" is used as a synonym for "processing time;" however, we distinguish between these terms for clarity and consistency with the operations research literature. Reducing cycle time is a core objective of lean manufacturing. Decisions and policies that lead to shorter cycle times are usually very beneficial for today's factory; those that lead to longer cycle times are not.

Variability has been called the universal enemy (Schonberger 1986). This may seem melodramatic, but in manufacturing it is actually true. (The variability to which we are referring is the variability in processing time.) Variability can be seen as anything that disrupts the flow. There are many reasons why it occurs in a production operation, some of which are listed here:

- unreliable equipment
- lack of standardized procedures
- uncontrolled environmental conditions
- long setup operations
- large production lots
- late deliveries from suppliers
- inappropriate management decisions.

Variability encumbers the factory. It leads to congestion, high inventory, long lead-times, and many other operational difficulties. Several researchers and practitioners have observed that the techniques usually associated with lean manufacturing are actually designed to reduce processing time variability (Inman 1993; Crawford and Cox 1991; Hopp, Spearman, and Woodruff 1990). Much of this book is devoted to explaining how today's factory can realize the benefits of reduced variability.

The importance of reducing variability is highlighted by the tremendous impact it has on the entire manufacturing operation. The following impressive list includes only some of the direct and immediate benefits of reducing variability:

- shorter cycle times
- shorter lead-times
- lower WIP
- faster response time
- lower cost
- greater production flexibility
- higher quality
- better customer service
- higher revenue
- higher throughput
- increased profit!

# A Science of Manufacturing

In 1992 two profound questions regarding the fundamental nature of manufacturing were posed to the operations research community:

Manufacturing systems are man-made artifacts. Is it possible, in these created worlds, to discover what might be called "laws of manufacturing?" If so, it can be argued, such laws would help establish intellectual foundations for a discipline of manufacturing. On the other hand, if such laws cannot be found, what other forms of knowledge will help us design, analyze and control better manufacturing systems? (J.D.C. Little 1992: 7).

We feel strongly that the answer to Little's first question is "yes, there are laws of manufacturing." Like the laws of physics, these laws can be discovered empirically, and they hold true for a remarkably wide range of situations. These laws constitute a *science of manufacturing*. They help clarify why production systems behave the way they do. Solutions to real problems in the factory can be developed deductively using these general laws. These solutions tend to be more effective and can be applied more confidently than solutions developed inductively from case studies and individual experience.

Little's second question concerns the existence of helpful knowledge other than the laws of manufacturing. Again, we feel strongly that the answer is yes. There is a *philosophy of manufacturing* that describes a holistic model of factory operations and production processes. This philosophy is known as *lean manufacturing*, and, when applied properly, it can lead to remarkable improvements in any factory's performance.

## Overview

American manufacturing has evolved in response to changes in the economic and technological environment. *Colonial craft production* flourished when markets were limited, products were highly customized, and advanced technology was unavailable. When demand suddenly increased and greatly outpaced production capacity, the *American System* emerged. Improvements in manufacturing technology made the dream of interchangeable parts a reality. This allowed the use of lower skilled labor and paved the way for vertical integration. It also enabled the concept of "build-to-stock." The benefits of the American System were amplified when mechanized conveyance and the moving assembly line led to mass production. The era of *mass production* was accompanied by almost unlimited customer demand allowing the factory to build-to-forecast since virtually everything produced could be sold.

How is the manufacturing environment different today? Today's customers demand low prices, perfect quality, and increasing variety. Today's factory has the capacity to produce far more products than it can possibly sell. Today's products are sometimes on the market for less time than it takes to design and produce them. Today's technology offers communication channels and factory automation that were science fiction only a few years ago. What are the implications for today's factory?

Just as craft production was ill suited for producing pistols at Colt's armory, the American System was insufficient for producing automobiles for the masses. Mass production, which has served industry so well for most of this century, is now causing serious difficulty for many companies that still cling to its tenets. Today a *different way of thinking* is gaining wide acceptance as an alternative and enlightened approach to manufacturing. This "different way" is known as *lean thinking*, and the production philosophy is known as *lean manufacturing*.

Lean provides some of the same benefits usually associated with mass production, such as high efficiency, and it can certainly be used in high-volume manufacturing. However, lean principles are profoundly different from mass production. Lean principles do not assume a constant, unlimited customer base. Lean manufacturing provides increased flexibility and responsiveness to changing customer demand. The lean approach to production control is extremely robust and efficient, even in the presence of high uncertainty and variability. Lean manufacturing provides simultaneously the highest quality at the lowest cost in the shortest amount of time.

## Running Today's Factory

*Running Today's Factory* presents a timeless methodology to help manufacturing managers run their factories as efficiently and as profitably as possible. The material presented is based on many years of manufacturing experience and the best scientific research available. We offer a coherent set of strategies that can be applied to virtually any "production" situation from growing citrus seedlings in a production flow greenhouse to producing electromechanical devices for use on board the orbiting space station.

It is unlikely that tomorrow's customer will become less demanding, so *today's factory must focus on satisfying the customer*. Satisfying the customer while making a profit requires the factory to be flexible, to produce perfect quality, and to do so as efficiently as possible.

It is also unlikely that tomorrow's competition will be less aggressive, so *today's factory must improve continually just to stay in the running*. When applied strategically, the principles and practices presented in this book will help today's factory gain and maintain a competitive advantage for today and tomorrow.

# Chapter 2

# The Manufacturing Challenge

## Introduction

Manufacturing companies have a tremendous economic impact on the communities in which they are located. Factories generate a tax base for local government. They employ local residents and provide trade for local businesses. Factories also purchase and use products and services, creating employment in other sectors of the economy. Directly and indirectly an estimated 31% of total economic activity in the U.S. can be attributed to manufacturing companies. "Millions of service-industry workers—including lawyers, consultants, truck drivers, and software developers—are employed because firms create a demand for their talents and skills" (Sheridan 1996: 21).

Manufacturing jobs are vitally important to local and state economies, and government programs, even at the national level, provide special technological and financial assistance to manufacturers. One such program is the Manufacturing Extension Partnership sponsored by the National Institute of Standards and Technology.

State and local governments understand the benefits of a strong manufacturing sector and often provide incentives to manufacturers who locate within their borders to stimulate economic growth. For example, the Arkansas Economic Development Commission promotes Advantage Arkansas to stimulate economic growth. The program offers "a job tax credit to qualified companies that locate within the state"(1998: 14). A publication from the South Carolina Department of Commerce describes a "valuable financial incentive offered to qualifying new and expanding industries in South Carolina that can be realized through Jobs Tax Credit." The Jobs Tax Credit provides a credit to corporate income tax or premium tax based upon new jobs created by a company locating or expanding a facility in any county in the state (1998: 6).

Alabama spent approximately $300 million to entice Mercedes to open a factory near Tuscaloosa. With an estimated workforce of 1,500, that equates to $200,000 per job. Other examples include the Toyota plant in Kentucky, the BMW plant in South Carolina, and the Saturn and Nissan plants in Tennessee (Table 2.1) (Myerson 1996). These states have determined that the positive economic impact of having these factories is worth hundreds of millions of dollars in incentives.

| State | Manufacturer | Year | Cost per Job |
|-------|-------------|------|-------------|
| Tennessee | Nissan | 1980 | $11,100 |
| Tennessee | Saturn | 1985 | $26,700 |
| Kentucky | Toyota | 1985 | $49,900 |
| South Carolina | BMW | 1993 | $65,000 |
| Alabama | Mercedes | 1996 | $200,000 |

*Table 2.1* State Incentives for Opening Local Factories

Employees expect the factory to provide a steady income, a secure future, and a place to refine skills. Employees may also see the factory as an opportunity for education and personal growth. Traditionally less than 20% of manufacturing jobs require higher education (Ferdows 1997), so employees may see the factory as a way to get ahead financially without the need for college. In spite of popular belief, manufacturing jobs *do* pay well. In 1993 manufacturing jobs paid an average of $12 per hour—higher than the average of any other industrial sector such as finance, retail trade, or wholesale trade. In 1995 the average weekly paycheck of a manufacturing worker was $512, compared to $369 for service industry workers and $221 for workers in the retail sector (Sheridan 1996: 20).

The owners or stockholders see the factory primarily as a moneymaking venture. They have a strong incentive to improve the financial performance of the factory. If production costs are lower and revenues are higher, then the factory generates more profit for the owners.

The factory's customers, whether individuals or businesses, expect to pay a fair price for a quality product that is delivered on time. In addition, customers may value innovation, quick delivery, customization, personal attention, service, or the ability to purchase items in small quantities. Other customer demands may include price reductions, just-in-time delivery schedules, certified compliance to standards and specifications, and involvement in the early stages of product design (Fleischer and

Liker 1997; Sandelands 1995; Kinni 1996). The demands of the customer have changed and evolved over the last 400 years, and much of this chapter is devoted to understanding this transition and how it affects today's factory.

It is not surprising that the expectations and needs of one group often contradict those of another. The customer's need for low price may oppose the employee's need for a high wage. Both may be at odds with the stockholder's objective of a high return on investment. It is easy to see that "what a factory needs to do" depends on one's perspective. Yet, each expectation of the factory has the same underlying requirement: the factory must be successful.

Without the financial success of the factory, the community does not benefit, the workers may be forced to find lower-paying jobs in other sectors, and the customers may have to find other manufacturing companies to supply their needs. Without the success of the factory, the stockholders lose their investments. The community, employees, customers, and shareholders all depend on the factory running successfully. *Today's factory is a complex association of diverse groups of people—all with their own interests—working together for success.*

This chapter lays the groundwork for thinking about manufacturing in a new way. We begin by defining the manufacturing challenge. Then we explore how factories met the challenge in the past. We also consider the social, economic, and technological context in which these successful approaches evolved. This retrospective provides insight into how the manufacturing challenge can be met today.

## Manufacturing Meets the Challenge

Manufacturers have been operating successfully in America for nearly 400 years. Artisan workshops produced handmade goods in the earliest American colonies, and technologically "advanced" factories have been operating here for 200 years. These early manufactories and factories faced many of the same challenges we face today. They had to procure raw material, convert it to a product, sell the product to a customer in a timely manner at a price the customer was willing to pay, and make a profit at the same time. As we will see, the way this challenge has been met has changed and evolved throughout the history of American manufacturing; however, the challenge has remained the same: *make what the customer wants when the customer wants it at a price the customer is willing to pay.*

## Artisans and Craft Production

Early settlers brought little with them other than the skills to make what they needed. A blacksmith named James Reed was among the original settlers of

Jamestown in 1607, and a cooper named John Alden came to Plymouth on the Mayflower in 1620. Every succeeding ship that came to the New World brought artisans and craftsmen (Tunis 1965). Men built houses and furniture from the timber they cut and hewed. Women made soap and candles, ground corn by hand, spun wool and linen, wove cloth, and sewed garments for their families to wear. As the early colonists became established, they began to rely on the more skillful members of their communities to "produce" certain items, and they paid for these items with surplus household products or services. The first manufactories emerged when the work of making things for others began to consume an individual's full time effort.

Southern colonists found abundant natural resources, fertile soil, and a favorable growing climate. This environment provided products and crops that could be exported to England for purchasing credit. In exchange for items such as tar, pine, tobacco, turpentine, rice, cotton, and indigo, they got much of what they needed *including manufactured items.* As a result, the Southern planters had little need for local artisans and craftsmen, and the basis for future manufacturing was not established to the extent that it was in the North.

Northern colonists did not have much to export other than timber, and the best of that was expropriated by the British Navy. Consequently, they had difficulty obtaining purchasing credit with which to buy English goods. They generally had to make what they needed from local resources. Artisans who came to the North found plenty of work (Tunis 1965). This set the stage for an American style of manufacturing that would later revolutionize industry.

From 1660 until the Revolutionary War, the British government restricted exports from America. Products approved for export were raw materials needed by British artisans, items needed by the British Navy, and, of course, tobacco, which does not grow well in England. For example, it was perfectly acceptable for American colonists to export pig iron smelted from American ore with American charcoal, but it was forbidden to export finished goods produced from the iron. Manufactured goods such as pots, pans, hinges, and tools were made in England by British craftsmen and offered for sale back to the colonists (Tunis 1965).

The master craftsman of the early 18th century faced considerable difficulty. He was the "purchasing department," locating scarce raw materials and negotiating prices. He was the "primary manufacturer," "supervisor," and "retailer." Products were often customized, so he was also the "design engineer." He was chronically short of help, and his living conditions were commonly overcrowded because he provided a home to everyone who worked in his shop. He also had to collect debts from his customers and deal with his creditors. Even though his situation was different, *the challenge for the master craftsman was the same as ours today: to make what the customer wants when the customer wants it at a price the customer is willing to pay.*

Let's summarize the conditions with which the master craftsman had to contend. His market was extremely limited because American craft products could not be exported to the vast European markets. His work was generally performed on a "bespoke" or "make-to-order" basis. Consequently the craftsman designed and produced custom products for specific customers, and the range of items produced was extensive. Even if demand had been greater, production rates were limited by the labor-intensive nature of the work and the absence of advanced manufacturing technology.

## The American System of Manufacturing

In the second half of the 18th century, two inventions revolutionized British manufacturing: the steam engine (1765) and a practical mechanism for transmitting its power (1781). These inventions spawned the *Industrial Revolution* paving the way for steam-powered transportation such as ships and trains, and industrial devices such as manufacturing and mining equipment. The availability of coal in England meant that inexpensive power was available in locations that were removed from the waterways. This liberated the factory from reliance on waterpower; steam engines could drive manufacturing machines that were technologically advanced by 18th century standards (Hopp and Spearman 1996).

These inventions were not available to the American colonists because England prohibited the exportation of technology until well past the Revolutionary War. Any attempts at importing machinery (or the patterns to make it) from England were thwarted by the British government. Technologically advanced manufacturing equipment did not even exist in America until the 1790s, and that was the result of some early industrial espionage.

The intrigue focuses on Samuel Slater, born in 1768 in Derbyshire, England. At the age of fourteen, he apprenticed in a textile manufactory. One of the owners was Richard Arkwright, the inventor of a technologically advanced spinning frame, which could spin many threads to varying degrees of fineness or hardness. During his apprenticeship, Slater gained a thorough knowledge of textile manufacturing in general and Arkwright's machinery specifically (Hopp and Spearman 1996).

American manufacturers tried repeatedly to invent technologically advanced textile machinery, but all efforts failed. In 1789 the Pennsylvania legislature even offered a reward for the "invention" of such equipment. Meanwhile, back in England, Slater heard about the bounty being offered in Pennsylvania, but he knew that technical drawings and models of textile machines could not be exported. Moreover, it was absolutely *forbidden* for anyone with such technical knowledge to emigrate to the United States.

Twenty-one-year-old Samuel Slater recognized an opportunity and departed for the New World disguised as a farmer (and without even telling his mother). There he met two important figures in the American textile industry, Moses Brown (for whom Brown University is named) and his son-in-law. With their financial backing and help from a group of skilled New England artisans, Slater constructed versions of Arkwright's machinery from his own detailed knowledge and eidetic memory. Together they established the first technologically advanced textile mill in the New World at Pawtucket, Rhode Island, in 1793.

So, it is clear that large-scale American manufacturing had its origins in British industry, but once established it evolved quite differently. American manufacturing took a distinctive form for two reasons. First, the primary energy source was different. Unlike England, America did not have a plentiful, affordable supply of coal. (The Pennsylvania coalfields would not open until the 1820s.) Consequently, early American industry did not benefit *significantly* from the invention of the steam engine. However, America did have abundant and widespread sources of waterpower. As American manufacturing expanded at the end of the 18th century, it relied heavily on the rivers.

Second, America did not have the strong tradition of craft guilds that were prevalent in England. In British textile manufacturing different stages of production (such as spinning, weaving, and dying) were performed by different people in separate occupations with distinct guild affiliations. These guilds had no incentive to simplify or centralize textile production. In contrast, American textiles had formerly been produced at home or by village craftsmen with no guild affiliation. Without the self-interested influence of the guilds, there was no resistance to centralizing manufacturing operations.

So American manufacturers took to the rivers and began consolidating disparate operations under one roof. This practice is known as *vertical integration*. Among the first to do this successfully was Francis Cabot Lowell in Massachusetts. In 1814 he combined all the processes required to make cotton cloth, and powered his equipment with water (Hopp and Spearman 1996: 20). Interestingly, the plans for Lowell's power loom had been smuggled out of England!

Meanwhile, an even more significant change was taking place. American manufacturers were pursuing the use of *interchangeable parts* in their assembly operations. This idea was not new. As far back as 1416, the Venetian arsenal used interchangeable parts in warships, and French manufacturers in 1785 produced muskets from interchangeable parts. For some reason, however, the idea did not catch on in Europe, and the practice was abandoned in favor of craft techniques. American manufacturers continued to develop the concept to its full potential.

In 1801 Eli Whitney demonstrated the concept of interchangeable parts to President-elect Thomas Jefferson and other officials in Washington, D.C. Components from disassembled muskets were put in piles, and randomly chosen parts were reassembled into complete muskets. The successful demonstration won Whitney a two-year contract to produce 10,000 muskets. The use of interchangeable parts not only permitted easy repair of finished guns, but allowed Whitney to set the almost inconceivable production goal of 5,000 muskets per year (Pursell 1981b).

Whitney's contract was essentially a test case for this production method, but production was interrupted constantly by unforeseen obstacles. Epidemics and shortages of parts delayed the completion of the contract for more than ten years. Paradoxically, the archeological and historical evidence indicates Whitney's guns were produced from parts "that were not even approximately interchangeable" (Shenkman 1994). Even if Whitney was more successful as a promoter than a practitioner of the interchangeable parts production method, his contribution cannot be overstated (Hounshell 1984).

Large-scale centralized factories and interchangeable parts made possible a totally new system of manufacturing. *This system involved producing discrete components on specialized equipment to exacting tolerances and then assembling them into finished products.* This uniquely American approach spread from the production of guns and clocks to sewing machines, typewriters, locks, bicycles, and many other manufacturing endeavors, eventually including automobiles.

The Crystal Palace exhibition in London in 1851 displayed many items produced by this *American System of Manufacturing*. Among them were sewing machines by Singer, "unpickable" padlocks by Alfred C. Hobbs, Samuel Colt's revolving pistols, interchangeable guns by Robbins & Lawrence, and a mechanical reaper by Cyrus McCormick. The British were so impressed that they sent industrial investigators to the United States to learn all they could about the American System (Wallechinsky and Wallace 1975; Pursell 1981b). By the mid-19th century, Eli Whitney's dream of a manufacturing system based on component interchangeability had become a reality, and America had become an industrial superpower.

Contrary to popular legend, the phrase *"American System of Manufacturing"* was not used at the 1851 Crystal Palace Exhibition. The phrase was introduced unambiguously in the ninth edition of the *Encyclopedia Britannica* in an article about locks. "The same system has been adopted in the government gun manufactories, and for clocks and watches; and no hand work can compete with it" (Hounshell 1984: 333).

Let's summarize the conditions that gave rise to the American System of Manufacturing. The market was expanding rapidly, the United States had become an independent country, and exports were no longer restricted. The manufacturer generally produced "to order," but the practice of stocking products and components became a viable alternative because machine-made products could be made in advance. One consequence of interchangeability was increasing homogeneity (less variety) of finished products. Production rates rose to unprecedented levels; only semi-skilled labor was required to manufacture most items, and the amount of labor required was greatly reduced due to the use of precision machinery. The advantages of the American System of Manufacturing were a high production rate, consistent quality, low production costs, and less skill and training of workers.

## Mass Production

In 1925 the American editor of the *Encyclopedia Britannica* wrote to Henry Ford asking him to write an article describing his production methodology. The article was actually written by Ford's spokesman, William J. Cameron, and first appeared on September 19, 1926 as a full-page feature in the Sunday edition of *The New York Times*. It was entitled "Henry Ford Expounds Mass Production: Calls It the Focussing of the Principles of Power, Economy, Continuity, and Speed." The basic principle of mass production, namely *economy of scale*, set the course of American manufacturing for most of the 20th century (Hounshell 1984).

While the automobile industry was not the only manufacturing sector to embrace mass production, the Ford Motor Company does provide an excellent case study of the transition from the American System to mass production. In the summer of 1906, Henry Ford hired Walter E. Flanders to be the overall production manager of the Ford Manufacturing Company and the Ford Motor Company. Flanders, a skillful machinist from Vermont, had learned about high-volume production and the American System while working at the Singer Manufacturing Company. Ford immediately grasped the advantages of the American System and gave Flanders full authority to transform the production operations. Henry Ford, like Eli Whitney, *promoted* the idea of interchangeability before he actually achieved it, but through the efforts of Flanders and an able colleague, Max F. Wollering, interchangeability of parts eventually became a reality.

One of Wollering's contributions was implementing standard fixtures and gauges so an unskilled worker could machine parts with the same quality as those produced by a skilled machinist. Standardizing the fixtures and gauges improved precision to a level at which true interchangeability was achievable. Precision was critical because even slight deviations from specification would necessitate hand fitting and adjustment in the assembly operation. High-volume manufacturing could only be achieved through high precision and high quality.

One of Flanders' accomplishments was to rearrange equipment sequentially in order of production operations. Before his arrival, machines had been grouped by function. For example, all milling machines were located in the milling department. Under Flanders' influence, machines of virtually any function were interspersed. A workpiece was moved from one machine to the next until it was finished. Flanders even went so far as to place furnaces in their proper sequential positions, so that when a part reached a stage in the production process requiring heat treatment, the part also arrived at the furnace! This close arrangement of machinery minimized material handling and prevented piles of work-in-process from accumulating between machines. Flanders and Wollering used dedicated and single-purpose machines to facilitate this sequential arrangement. These changes might have been revolutionary in automotive production, but they were hardly new to American industry. After all, Flanders and Wollering were simply applying the tenets of the American System of Manufacturing to a *brand new* industry.

Mass production has its roots firmly planted in the American System; however, the transition to mass production required incorporating two more practices into Ford's fledgling manufacturing system: *mechanized conveyance* and *assembly line production.*

Mechanized conveyance provided fast, efficient transfer of material from one processing station to the next. Its first application in American industry was in 1784 in a flourmill outside Philadelphia, Pennsylvania. Oliver Evans linked the various processes of flour milling and refining with chutes, conveyors, and elevators. Using waterpower to drive the conveyors, grain moved automatically through the mill. Grain entered the mill and was transferred from one process to the next until it emerged as finished flour. Evans estimated that the system reduced flour production costs by 50%. The system spread quickly throughout the American flour milling and brewing industries. Less than one hundred years later in 1873 American meat processing plants were using both conveyors and (dis)assembly lines. Ford knew about these practices and used them later as models for his automotive manufacturing operations.

The Model T is a classic illustration of the transition to mass production in American manufacturing. Ford introduced the Model T in 1908. One year later, he announced it was the *only* model that Ford Motor Company would produce. All variations such as the runabout, touring car, town car, and delivery car would use an identical chassis. Even though the Model T was originally painted green with a red stripe (Shenkman, 1994), the "any color so long as it's black" statement characterizes the single-model philosophy of mass production.

Ford's new factory in Highland Park opened in 1910 to accommodate the increase in Model T production. By that time, Flanders had left Ford, but his influence was clearly evident in the Highland Park facility; the new plant used special single-purpose machinery and sequential layout. During the first few years of operation, the Highland Park factory made only modest gains in productivity as engineers experimented with different production techniques.

The flywheel magneto assembly department serves to illustrate Ford's transition to mass production. Originally the magneto assembly employees worked at individual benches making complete flywheel assemblies from a myriad of component parts. Using the individual assembly method, each worker produced between 35 and 40 assemblies per day at a rate of *one every 20 minutes.*

On April 1, 1913, they came to work unaware that they were about to make manufacturing history. They found that the various components were no longer placed at their individual workbenches. Instead, the components were placed beside a sliding "assembly line." The foreman explained that each worker was to perform a minor task, such as installing a magnet or starting some nuts, and then send the assembly down the line to the next worker. Using the assembly line approach, the workers *together* produced 1188 assemblies at a rate of *one every 13 minutes* per worker. Within a year, the productivity per worker increased by a factor of four to *one assembly every five minutes per worker* (Hounshell 1984).

The first chassis assembly line in August 1913 was crude: it literally was pulled by a rope. However, it was a phenomenal success! By the end of April 1914, just one short year after Ford's first experiments with the assembly line, the labor involved in chassis assembly had decreased from 750 man-minutes per car to 93 man-minutes per car—an 88% improvement! Eventually, Ford was able to pro-

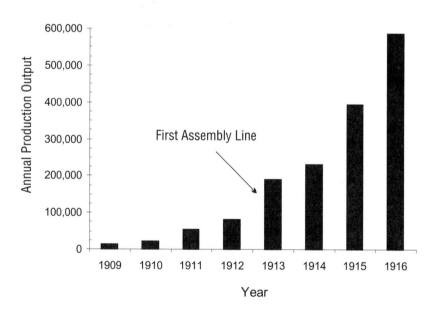

***Figure 2.1*** *Model T Production (1909–1916)*

Source: *Data from David A. Hounshell,* From the American System to Mass Production, 1800-1932. *(Baltimore: The Johns Hopkins University Press, 1984), 224.*

duce a complete Model T every 10 seconds (Flink 1981). A productivity improvement on this order of magnitude distinguishes mass production from all previous manufacturing systems. Figure 2.1 shows annual production figures for the Model T from 1909 to 1916. Notice how sharply the production increased after Ford began experimenting with the assembly line in 1913.

Ford believed that the Model T was the ideal automobile. He had developed the car for the masses, and he wanted every American to be able to afford one. Enormous "economies of scale" were achieved by focusing production efforts exclusively on this vehicle. From 1909 to 1916, the purchase price of the Model T fell from $950 to $360, and by 1921 Ford had 55% of the automotive market share.

The popular legend is that the Model T remained unchanged from its introduction until its cancellation in 1927. This is not entirely true. Minor changes were made to its design, but these modifications did not satisfy the customers' desires and expectations. Market share fell from 55% in 1921 to 30% in 1926. Financial losses grew from $30 million in 1927 to $70 million in 1928 while Ford struggled to introduce the new Model A. By 1936 Ford's market share had dwindled to 22% (Hounshell 1984).

This period of Ford Motor Company history highlights the primary characteristics that, even today, epitomize mass production. Mass production is characterized by *extremely high-volume production of a single or limited number of products.* Usually the item produced has a limited number of options and cannot be customized extensively by the factory. Often the production process involves automated conveyance and numerous processing stations, each performing only a fraction of the tasks required to complete the item.

## Today

Throughout the 20th century, management has embraced the mass production philosophy, and factories have realized unprecedented gains in efficiency and productivity. The customer, however, has not fared so well. The same "standardized" and "interchangeable" tenets that are the foundation of mass production have fundamentally changed the relationship between the factory and the customer. *Mass production insulates the factory from variability in customer demand, but it does so by categorizing customers and products and trivializing customer interaction* (Lampel and Mintzberg 1996).

By the 1920s Americans had accepted the standardization required for successful mass production in exchange for the availability and affordability of many commodities. A 1929 survey of 84 product classes shows *reductions in variety* as high as 98%. For example, bed blankets, which came in 78 different sizes in 1921, were available in only 12 different sizes just eight short years later; hospital beds went from 33 different varieties to one "standardized" bed; and, of course, we could all bathe quite adequately in those "standard" five-foot tubs. Men's hats, which had come in 100 different colors in 1921, were available in only nine colors in 1929. The 23,560 custom shoemaking establishments of 1900 had been reduced to just a few hundred in the early 1920s (Lampel and Mintzberg 1996).

As we stand on the brink of the 21st century, let's look at what the customer is demanding today. From 1980 to 1990, the variety of breakfast cereal being sold grew from 88 to 205 (Womack 1993). Standard rotary telephone service has been augmented by many custom features such as touch-tone service, call forwarding, call waiting, caller ID, caller ID blocking, "busy buster," voice "mailboxes," and conference calling, not to mention cellular and digital phones. Music stores are offering "customized" cassette tapes and CDs that are produced in a matter of moments with the precise musical selections specified by the customer. Book publishers are offering personalized children's and business books. Shoppers can "manufacture" their own greeting cards in grocery stores, malls, on their personal computers, and via the Internet.

The National Bicycle Industrial Company in Japan builds customized bicycles in mass quantity. Storekeepers record the customer's choice of size, color, and design; then the customer is measured on a special frame. A computer generates the engineering drawings, and robots measure, weld, and paint the frame. Skilled workers add the finishing touches including the customer's name applied by a silkscreen process (Maital 1991). Personal computer manufacturers such as Gateway 2000 and Dell have been immensely successful producing customized computers for schools, businesses, government, and individual users on a make-to-order basis (Serwer 1997). IBM, Compaq, Packard Bell, and Hewlett-Packard are trying to adopt similar strategies (Carlton 1997; Kirkpatrick 1997).

Switzerland's largest watch manufacturer, ETA SA, produces the Swatch in an amazing array of colors and styles. (Have you ever even seen two Swatch watches that are exactly the same?) Custom Foot, a pioneer in the shoe industry, keeps no inventory. Customers choose color, type of leather, and style from about 100 display models. Then precise measurements are made, the order is sent by modem to Florence, Italy, and the shoes are produced in a factory. Within three weeks, the shoes are ready at a price that is about 80% less than traditional hand-made shoes. How many of us have purchased a personalized T-shirt from our favorite vacation spot? Levi-Strauss offers "Personal Pair" jeans for women made to their exact measurements. These jeans were such a success that Levi's now offers a similar product for men (Martin 1997). So much for the chocolate truffle industry!

The point is not that these goods and services are customized. Certain traditional "crafts" such as fine jewelry making, personal tailoring, eyeglass lens grinding, and fine restaurant fare have always been customized. The point is that these companies are offering their customized products and services in *mass production* quantities. *Mass customization* is a growing trend in business today, and these examples raise a question. In light of all these changes in customer expectations, does it make sense to run today's factory the same way it ran in the past?

## What Has Really Changed?

Certainly, the market demand has grown substantially, driven by a growing population, a rising standard of living, and an explosion in communication and transportation technology. However, factory output has not only kept pace but has *exceeded* demand in many sectors! For example, automotive production capacity outpaced demand by 18.5 million units in 1997, and projections for the year 2002 show the expected excess capacity to exceed 20 million units (*The Wall Street Journal*, 2 March 1998: A2).

Today's customers are much more discriminating. More choices are available, so consumers are getting more particular. Customers no longer make purchasing decisions based solely on product attributes or cost. Speedy delivery and high quality are now just standard requirements for doing business. Today's customers want even more! They want products made exclusively for them. From Levi's jeans to personal computers, customers are demanding "customized" products made to their precise specifications and delivered on short notice. In fact, the computer system on which this book was written was custom-made to the authors' specifications and delivered in three business days. Today's factory *must* be cognizant of these changes in customer demands and develop appropriate manufacturing strategies to remain competitive.

Technological advances in machine tools are providing speed and precision that would have been considered science fiction only 20 years ago (DeJong 1998). Computer-aided design and computer-aided manufacturing are compressing dramatically the time from concept to product, and we can now grow three-dimensional prototypes out of vats of liquid plastic. In some cases these laser-generated "prototypes" can be used in actual applications. This responsiveness to individual consumer demand is reminiscent of the artisan's workshop where an early American colonist could order a product handcrafted to his specific needs. The factory's role has come full circle.

## Summary

In each period of American manufacturing, the factory faced different challenges, and manufacturing has evolved through the years to meet the particular needs of the times. Each style of manufacturing represents a successful approach, providing products for customers and profits for owners. Each production system (craft production, the American System, and mass production) is unique; yet, there are some common elements and techniques. While these elements may be similar, often they have played very different roles in the success of each system. Highlighting these similarities and differences helps us understand why manufacturing systems changed, and provides insight about how factories should run today.

Consider the factory's approach to *quality*. For the craft producer, quality had to be good enough to satisfy the customer. Quality problems that occurred during production could be corrected through hand fitting, which is a defining characteristic of craft production. In the American System, quality became much more of an issue. Parts interchangeability requires that each component be just like the next. This level of quality was sometimes assured through the use of piece-rate systems that only paid a worker for acceptable items. In the early days of mass production, high quality became even more important. High quality of components was a requirement for assembly line production, and high quality in finished goods was a competitive advantage.

Today quality takes on an entirely different dimension. Excellent quality in finished products is now simply expected as a matter of course. Excellent quality of purchased items and produced components is also required to support the low WIP, tight schedules, and the flexibility of today's factory.

Consider the factory's approach to *layout*. In the American System, the cost of riverfront property, the use of waterpower, and the need for natural lighting often prescribed the layout. Nevertheless, factory managers were aware of the inefficiencies of poor layout and tried to promote flow and minimize the cost and labor associated with material handling. In the early days of mass production, the constraints on layout were relaxed, and factories were able to adopt product-oriented or sequential layout.

> "There is another advantage in placing the machines in accordance with the sequence of operations; even though some machines are not worked to their full capacity the amount invested in them is well paid for from the fact that it is not necessary to carry nearly so much stock as when the machines are grouped according to their classification. Each group or department alone would in that case need to have nearly as much raw stock to work with as is necessary with this method to operate the whole series of machines to complete the part" (Bornholdt 1913).

This statement could have been made about a contemporary "lean" factory operating with single-piece flow manufacturing. However, the statement refers to Ford's early practice of arranging equipment in accordance with the sequence of operations. It is ironic that Ford later abandoned the practice in favor of process-oriented layout and process villages.

A natural extension of sequential layout is cellular manufacturing. While this is considered a signature practice of lean production systems, Ford implemented U-shaped layouts and manufacturing cells in the late 1920s.

> "On one of these multi-spindle or turret machines there are usually from five to seven stations in addition to the loading point. They are arranged in a circle so that the forging, after it has been machined at the different points as required, will complete the circle, returning to the operator as finished part. He removes it and places it on a conveyor, at the same time taking from the conveyor a rough forging" (Ford 1931: 130-31).

Early mass producers favored this arrangement because it minimized the material handling effort and maximized productivity and efficiency. Today's factory also orients equipment this way but for a somewhat different reason. Sequential layout provides the shortest possible cycle time while reducing costs and increasing flexibility and responsiveness to customer demand.

The ways in which manufacturing companies have used *human resources* are also quite interesting. Obviously craft production required a very high skill level. An apprenticeship of several years was required to learn all the tasks involved in making a product. In contrast, production under the American System required far less skill, and jobs were generally more monotonous; yet, many different jobs were performed under the same roof. Cross-training both relieved the monotony and gave the comparably small factory versatility and flexibility. Mass production required less skill because machines began performing even more of the precision or "skilled" work.

Today's factory considers its employees a very valuable asset. In general, the skill required is only moderate, but experienced workers who are cross-trained are especially valued because they can support a flexible production mode. The competitive advantage that multi-skilled workers brought to the factory under the American System was the increased adaptability of a small operation. The competitive advantage that multi-skilled workers bring to today's factory is the flexibility to respond to changing customer demand.

Consider also the way a factory uses *inventory*. The purpose of inventory has not changed over time: inventory buffers the factory against variability. This variability can be caused by interruptions in communication and transportation and disruptions of the production process (machine downtime, poor raw material, un-available components, long setup, worn tooling, and absenteeism). Variability can also be caused by fluctuations in customer demand. Whenever variability is present, *the factory must buffer or protect itself* from it. This buffer can take the form of excess capacity, extra time, or inventory. Traditionally the buffer of choice in American manufacturing has been inventory.

Under the American System, inventory was used to sustain a high level of pro-ductivity at a time when communication and transportation were unreliable. By the early days of mass production, communication and transportation had improved, and those factories producing a limited product mix could arrange with a reliable supply network to provide material just in time.

> "We have found in buying materials that it is not worthwhile to buy for other than immediate needs. We buy only enough to fit into the plan of production, taking into consideration the state of transportation at the time. If transportation were perfect and an even flow of materials could be assured, it would not be necessary to carry any stock whatsoever" (Ford and Crowther 1924: 143).

This approach to inventory was possible when Ford was making only one prod-uct, but when the company's operations expanded to include multiple models, this approach became too complicated.

Inventory serves the same purpose for today's factory that it served in the past. Recalling the three buffers mentioned above, using either inventory or excess capacity requires capital investment, which management is often reluctant to subsidize. The third buffer—time—can "cost" the factory its customers because customers today have high expectations for timely delivery. Running today's fac-tory successfully requires more than choosing the least detrimental buffer. It re-quires going one step further and reducing the need for buffers at all, in other words, reducing variability. Much of this book is devoted to understanding the nature of variability and how we can reduce it throughout the value stream.

*What has really changed?* This book is about running today's factory, and we now answer this question from the factory's point of view. In fact, *nothing has really changed*. Running today's factory requires meeting the same challenge: to convert raw materials into products that are offered for sale to customers on a timely basis and to do this profitably.

Table 2.2 recapitulates many socioeconomic and technical changes that have influenced American manufacturing during the past 400 years.

| Factory Aspect | Craft Production | The American System | Mass Production | Today's Factory |
|---|---|---|---|---|
| *Products* | Wide variety or customized | Uniform | Uniform | Wide variety or customized |
| *Production Control* | Make to order (Bespoke work) | Make to order and forecast | Make to forecast | Make to customer demand |
| *Production Technology* | Skilled artisans | Precision machinery<br>Semi-skilled labor | Precision machinery<br>Semi-skilled labor | Computer-controlled, high-precision machinery<br>Semi-skilled labor |
| *Production Methodology* | Hand fitted and fabricated | Interchangeable parts<br>Machines<br>Hand labor | Interchangeable parts<br>Automated machines<br>Hand labor<br>Mechanized conveyance | Interchangeable parts<br>Automated machines<br>Robots<br>Hand labor |
| *Market Demand* | Extremely limited | Market exceeds production capacity | Market exceeds production capacity | Market significantly less than production capacity |
| *Customer Requirements* | Availability | Availability<br>Product attributes | Availability<br>Product attributes<br>Cost | Product attributes<br>Cost<br>Quality<br>Customization<br>Speedy delivery<br>Innovation, *etc.* |
| *Information Technology* | Spoken word | Plans and drawings | Plans and drawings<br>Computers | Plans and drawings<br>Computers<br>Internet |

*Table 2.2 Changes that Influenced American Manufacturing*

# Chapter 3
# What's So Tough About Running a Factory?

## Introduction

We are all familiar with the everyday problems facing today's factory. Issues such as worker absenteeism, unreliable equipment, late deliveries, poor quality materials, schedule changes, and long setups certainly add to the challenges of running a factory. Sometimes systems and procedures are put in place to help the factory deal with these common problems. Often very creative people are put in place to work around dysfunctional systems and to make sure the factory runs properly in spite of the problems. The following story illustrates how creative people are able to run a factory in spite of problems and uncertainty.

We were recently holding a factory demonstration to contrast a *push production control system* with a *pull system*. In the push phase of the demonstration, toy trucks in three colors are built according to a forecast of customer demand. Trucks that are built but not "bought" by the customer are stored in a finished goods storage area. The forecast is accurate (on average), but there is enough variability in customer demand that there is always an accumulation of trucks of one color and a shortage of trucks of the other colors. The customer is dissatisfied because so many orders cannot be filled. In the pull phase of the demonstration, trucks are built only to replenish a marketplace of finished trucks. As might be expected, the pull system replenishes trucks at the same rate at which they are "bought," and, therefore, always results in a satisfied customer.

On this particular day, we asked the materials manager of the factory to play the role of the material handler in the demonstration. The push phase of the demonstration had an early stock-out as we expected. However, from that point on, the toy truck factory was somehow able to produce *exactly* what the customer needed. We were perplexed since we had developed the demonstration specifically to illustrate the opposite: that building to forecast results in a buildup of products the customer doesn't want and a shortage of products the customer does want. Several minutes later we realized that the materials manager was not following the delivery schedule. She was not supplying the truck factory based on the forecast but based on what the customer was actually "buying." She was performing the same role she performs daily in her "real" job. She was working around a dysfunctional system to somehow keep the factory supplied and running smoothly to meet customer demand. Her approach in the demonstration and in her real job is essentially an informal and self-taught pull system (Chapter 9).

Months later we found out that this materials manager had been concerned that the aspects of her job that she considered "creative" and "interesting" would disappear after the factory successfully implemented a pull system. In fact, the mundane and aggravating aspects of her job did disappear, and she was able to use her creativity and factory knowledge to adapt the pull system concepts to accommodate the specific needs of the factory.

Why is running a factory so tough? This chapter does not address the myriad problems associated with factory management. Instead, it focuses on solutions for today's factory and why they are so difficult to find and to implement. Several popular approaches are introduced, and we explain why their usefulness is often limited. The mathematical complexity of factory management is illustrated with a discussion of job shop scheduling and two detailed examples. We conclude this chapter by introducing two holistic approaches to factory management: lean manufacturing and factory physics.

## Diverse Expectations

The manufacturing challenge presented in Chapter 2 is "*make what the customer wants when the customer wants it at a price the customer is willing to pay.*" The factory has to meet this challenge and, at the same time, satisfy the diverse needs of the owners, workers, suppliers, and the community. These stakeholders have different ideas about what purpose the factory serves, and their expectations are sometimes incompatible. For example, the community needs a lucrative tax base and jobs for its residents, and the employees want a living wage and a chance for personal growth. The owners need a profitable return on their investment, and the customers expect a variety of products of perfect quality delivered quickly at an affordable price. Even if the factory had no operational problems, trying to satisfy so many expectations would make running a factory tough. The job of a manufacturing manager can sometimes seem like that of a circus acrobat: amaze and delight the audience and fulfill everyone's expectations while defying the laws of physics and nature.

## Conflicting Demands

Sometimes the operational demands on the factory are contradictory. Reconsider the manufacturing challenge: the customer's desire for high quality, low price, and speedy delivery. According to traditional manufacturing wisdom these demands are incompatible. For many years, American manufacturers have taken it for granted that high quality could be achieved only at great expense. "It used to be axiomatic that higher levels of quality meant higher costs" (Spira and Pine 1993: 26). This belief is still widely held. A statistical survey of 110 small manufacturers from a variety of industries reveals that *most American companies are actually willing to sacrifice quality* if it means they can achieve cost reductions, on-time delivery, and reductions in lead time and setup time (Temponi and Pandya 1995: 26-7). In other words, achieving low cost is considered more important than providing the customer with high quality. The predominant view is that high quality and low price are still mutually exclusive.

Table 3.1 shows several *traditional approaches* to meeting the manufacturing challenge. Notice how they conflict with one another. In the traditional view, *short lead time* is achieved by keeping excess inventory or having excess production capacity that can be tapped when needed. This is inconsistent with the low inventory and high machine utilization traditionally associated with low price. Another example is *high quality*. Traditionally a lower processing rate is assumed to yield higher quality whether the process is manual or automated. This conflicts directly with the higher processing rates traditionally used to attain short lead time and low price. A third example is *low price*. In the past, "economical" lot size has been used to reduce cost through economies of scale. This is in opposition to the small lot size necessary to support shorter lead time.

| Short Lead Time | High Quality | Low Price |
|---|---|---|
| Small lot size | Large lot size | "Economical" lot size |
| High processing rate | Low processing rate | High processing rate |
| High inventory | High inventory | Low inventory |
| Excess capacity | 100% inspection | Long lead time |
|  | Sorting out defects | Low quality |
|  | SPC | High utilization |

*Table 3.1 Traditional Approaches to the Manufacturing Challenge*

Do these three customer requirements intrinsically conflict with one another? There is growing industry evidence that indicates the answer is no. Short lead time, high quality, and low price do not necessarily conflict with one another. In fact, they are quite compatible and can be achieved simultaneously using mutually supportive methods. Producing high quality goods with a short lead time can actually be the most efficient and profitable way to run today's factory. How is this possible?

In the following chapters, we answer this question by addressing the *underlying causes* of poor quality, long lead times, and high cost. We present sound principles and practices that help today's factory achieve these three objectives: high quality, short lead time, *and* low cost.

## Anachronistic Dogma

Today's uncertain manufacturing environment changes rapidly. Often the time required to obtain market data and to determine expected product demand is longer than the total life of a product! So, the factory management must determine material and capacity requirements, and must plan and schedule production based on incomplete or inaccurate data. This uncertainty adds to the difficulties of running the factory. Solutions used in the past are often ineffective in today's turbulent manufacturing environment.

Tried and true manufacturing strategies should not be rejected without good reason. However, as demonstrated in Chapter 2, we should not follow blindly the management practices of the past. They should be challenged and scrutinized to assess their relevance today. Strategies that are still valid should be retained. Those that are obsolete should be replaced by more appropriate approaches. For example, electronic data transfer, overnight delivery, and the Internet provide new solutions to the old problems of materials management. If a supplier can access the factory's inventory and production information in real time via the Internet and deliver goods overnight, there is little reason for placing large orders based on forecasts and long lead times.

Outdated beliefs and attitudes lock the factory in the past and prevent people from looking for better, more appropriate solutions. "But we've always done it this way." "If it ain't broke, don't fix it." "Quality costs too much." "Maximize the economies of scale." "Nothing ever changes." These statements reveal that the speaker is trapped in the past, and these attitudes prevent the factory from adopting creative, new solutions to today's problems.

Let's consider a very successful product that dominates today's laundry detergent market and has for many years. Should that product be improved even though it is doing very well? If not, when *should* the product be changed? After sales begin to decline? Tide has been the leading laundry detergent since the 1940s when it was introduced. It has undergone 8,000 reformulations: new packaging and new ingredients. Was Tide ever "broke?" Did it ever require "fixing?" Regardless of the way it was "always done," Procter and Gamble continues to invent, improve, and respond to the changing, competitive environment. This progressive business strategy has kept Tide sales ahead of all challengers.

We recently attended a meeting at a plant that is attempting to adopt lean manufacturing. This company has invested millions of dollars to develop a coherent production system based on lean principles and patterned after the Toyota production system. The plant manager, who resisted the transformation, sternly lectured his operating committee that "we already have a ____ production system," meaning that the company had been in operation for nearly a century and obviously knew how to make its product! He insisted that his departments be rewarded for producing more than needed by the next process and that the plant be rewarded for producing more product than the customer wanted to buy. These attitudes lock the company in the past, do not encourage improvement, and effectively prevent the plant from making any significant advances.

Consider another example: the Compaq Computer Corporation. In July 1997 Compaq announced an official policy of building personal computers to customer order. Eight months later in March 1998, Compaq was having its worst quarter since 1991. Compaq was in danger of losing money and was hoping to just break even. Why? Swollen inventories were clogging store shelves. Dealers were giving away monitors to entice buyers. How could this happen when, just a few months earlier, the company had adopted an official policy of building to customer order? Earl Mason, CFO of Compaq Computer Corporation, has the answer: "A product guy will say, 'I know these orders are coming. Let's go ahead and build it rather than close the line' " (Ramstad 1998: B1). Horatio Lonsdale-Hands, Chairman of ZuZu, a Mexican food franchise based in Dallas, calls this the "unskilled mentality: let's make it all now, and then we won't have to make it during the rest of the day" (Tannenbaum 1997).

Compare this with the business strategy of Dell Computer Corporation. "The company operates with a bare minimum of inventory. Since it builds a computer after an order is placed, it generates substantial free cash even as it has maintained a nearly 50% annual-revenue growth rate" (McWilliams 1999). How has this strategy affected company performance? Dell came in first among one-thousand companies on the *The Wall Street Journal's* Shareholder Scoreboard for return to investors over the past three, five, and ten years. For example, $1,000 invested in Dell in 1988 would have grown to over $300,000 by the end of 1998. Clearly a progressive way of thinking about manufacturing leads to very strong company performance!

## Inconsistent Goals and Performance Measures

Another reason running a factory is tough is that the business goals of the owners and the personal goals of the executives do not necessarily support the long-term success of the factory. Consider the business goal of increasing share price 35% by year's end or achieving a 25% before-tax profit. These goals are very reasonable, but do they foster behavior that leads to long-term competitiveness of the factory? Short-term goals and financial measures often lead to short-term actions. For example, one of the easiest ways to improve short-term financial performance is to reduce cost and cut back on capital spending. Investing in research and product development and building internal manufacturing capability do not pay dividends in the following quarter. They are investments for the future, so they are not encouraged by short-term financial goals.

We are familiar with a struggling manufacturing company that has been sold eight times in ten years. Each corporate owner had the same basic strategy: maximize the short-term gains and sell the company for a profit. Little or no investment was made in modernizing or replacing faulty equipment, much of which is so old that replacement parts cannot be procured. The short-term financial goals of the parent corporations did not support the long-term strategic positioning of the company. In the American manufacturing community, this is more common than we would like to admit.

In the 1950s Japan began funneling its best and brightest into engineering and onto the shop floor to develop manufacturing excellence and bolster its industrial viability. Meanwhile, America was developing a different approach toward business management and corporate competitiveness. This new approach was known as financial management. During WWII, a group known collectively as the Whiz Kids made an impression in the Air Corps for applying analytical techniques to managerial problems. After the war, Ford hired the Whiz Kids to develop a systematic way to manage the company. The Whiz Kid approach put exclusive emphasis on financial criteria. In the management system they developed, financial controls dominated every other factor of the business such as marketing, engineering, and manufacturing (Nicholas 1998: 16). Financial management did put accounting controls into place, but it had an unfortunate side effect. "The emphasis on marketing and finance long ago took top management out of the loop as far as operations were concerned" (Hopp and Spearman 1996: 169). Driving up share price has become more important than making a good product at a profit and developing manufacturing capabilities for the future. A former Chrysler Corporation executive vice president for manufacturing observes, "The prime interest of a money manager is not to create new products or to develop a competitive workforce, or to build innovative production processes. Their main interest is to do whatever it takes to drive up the company's stock price" (Sharf 1999: 21).

Sadly, the Whiz Kid approach is still alive and kicking today. It is taught in business schools and practiced in boardrooms. Cost-cutting or cost-avoidance measures are pursued with zeal in many companies, and closing plants to move manufacturing overseas has become a common item on many corporate agendas. For example, Levi Strauss & Company, the San Francisco-based clothing manufacturer, recently announced that it would lay off nearly 6,000 workers, close eleven plants, and shift manufacturing overseas *to cut costs* (*The New York Times*, 28 February 1999).

## Canned Solutions

There are cultural beliefs and attitudes that contribute to the American fascination with canned solutions. The American belief that there is a "best way" or a "right solution" is clearly evident in the way we perceive problems. Many managers see their factories' problems as straightforward issues that can be solved with an all-purpose, narrowly defined solution (Gupta and Somers 1993: 87). This attitude is rooted deeply in American scientific reductionism. In 1958 the economist John Kenneth Galbraith stated that we had solved the problem of production, and, therefore, we could move on to other things (Galbraith 1958; Hopp and Spearman 1996: 182). *This should be surprising and welcome news to any factory manager!*

Another relevant characteristic of American culture is impatience. Not only do we want the *best* solution, we want it *right away*. Researchers from Disney University point out that Americans have a tendency to live in the present. We are impatient and prefer action to planning. We are always angling for a shortcut (Hammond and Morrison 1996: 146). Hopp and Spearman have observed that "we Americans seem to have a resolute faith in a swift and permanent resolution of the manufacturing problem" (1996: 182).

So, couple our sincere belief in an ultimate "best way" with our impatience, and we are set up perfectly for the "quick fix" or the "canned solution." Shapiro calls attention to our propensity for fad solutions in her recent book, *Fad Surfing in the Boardroom: Reclaiming the Courage to Manage in the Age of Instant Answers* (1995).

Production managers are bombarded with a wide variety of slogans, fads, and theories for improving their manufacturing operations. Most have catchy *Three Letter Initials* (TLI), such as MRP, ERP, FMS, AMT, BPR, JIT, SPC, CIM, TOC, and TQM. Others, such as MRP-II, JIT-II, ZI, 5S, and 6-Sigma, also offer *the* answer to our manufacturing woes. Some of these approaches have developed a cult-like status, complete with charismatic champions and zealous followers.

Of course, there are no generic solutions that can be applied in all manufacturing situations. Even if a magic solution were found today, continual improvement would still be necessary to keep the competition at bay. Nevertheless, we in the manufacturing community continue to look for a silver bullet to solve the problems in our factories. This is dangerous. "People who want quick results too easily believe people who promise them" (Stewart 1995: 162).

Popular fads are not panaceas. Sometimes they may produce dramatic results, but they can also produce disastrous results (Cooney 1995: 95). Let's take a brief look at MRP, JIT, SPC, and TOC in order to point out their limitations and explain why they cannot lead to holistic solutions.

The original purpose of MRP (material requirements planning) was to determine what materials and components are needed and when (Orlicky 1975: 152). MRP is well suited for this because planning material requirements is a deterministic problem that is isolated from the actual circumstances in the plant. This strength is also its weakness when MRP is inappropriately used as a production scheduling mechanism. For this purpose MRP has been far less successful (Chapter 9) and has even been called the "100 billion dollar mistake" (Whiteside and Arbose 1984). MRP assumes a *fixed and known* lead time, while actual, "real factory" lead time is *variable and unknown*. Prudent users enter inflated lead times into the MRP system, and, consequently, orders are released to suppliers and to the shop floor sooner than needed. This, in turn, causes more congestion and longer lead times in the factory. When these longer lead times are subsequently entered into the system, the problems are compounded. This spiral continues, and the factory eventually becomes mired in its own WIP (Chapter 9).

JIT, just-in-time, has not fared much better as a stand-alone approach. To some people, JIT means just-in-time delivery, implying that suppliers hold excess inventory for delivery to the factory at a moment's notice. To other people, JIT has come to mean *anything that is good about manufacturing* (Spearman and Zazanis 1992: 521). Unfortunately, the term has acquired so many connotations that it has become almost meaningless. Our factory experience and evidence from industry indicate that *the JIT techniques themselves do not confer any significant competitive advantage*. They can contribute tremendously to a holistic improvement plan, but JIT techniques themselves do not constitute a complete solution (Chapter 7).

SPC, statistical process control, is often used to monitor and control manufacturing processes. Surprisingly, more than 70% of industrial processes do not satisfy the underlying assumptions on which traditional SPC is based. In these cases, Shewhart-style SPC charts are invalid. Recent advances in SPC methodology have broadened the applications for which it is valid and appropriate. Nevertheless, these advances remain largely a curiosity of academia and are rarely used in actual factories. More importantly, SPC is not specifically designed to be a tool for improvement; it is a tool for maintaining the *status quo* (Chapter 10).

TOC, the popular *theory of constraints*, has helped many factories improve their operations. In simplest terms, TOC involves determining which production process is constraining the overall operation and then improving that process. The constraint operation, or bottleneck, must be protected with extra work-in-process inventory (WIP) so it does not "starve" or shut down due to a lack of production material. TOC emphasizes the importance of scheduling, the importance of minimizing cycle time, and the folly of maximizing resource utilization. (In spite of these contributions, TOC has been criticized in academic literature, and support from the operations research community has been minimal. Some researchers even contend that the theory of constraints is not a theory at all (Spearman 1997: 29).)

One serious problem when applying TOC is that it can be difficult to identify the constraint operation. Often the bottleneck is not obvious. The most straightforward approach is to look for excessive WIP, but there are many other reasons why WIP might accumulate between processes. For example, a performance measure might encourage production of items regardless of whether the downstream operation or the customer is ready for them. In this case, a performance measure, rather than a bottleneck operation, is responsible for the buildup of WIP. Variability (Chapter 5) can also lead to excessive WIP. If a process has periods of rapid production followed by slow production or downtime, WIP accumulates. If the capacity utilization is high, the factory cannot recover from the slow production periods (Chapter 12). This results in *longer* lead times and even *higher* WIP levels. In any case, using WIP to identify the bottleneck operation can lead to erroneous conclusions.

Another difficulty identifying the bottleneck is that it rarely stays put; bottlenecks tend to wander around the factory. What causes this phenomenon? Bottleneck wandering is often attributed to product mix. Each product presents a different set of challenges, and the bottleneck for one may not be the bottleneck for another. Current research indicates another, more serious, cause of bottleneck wandering: disruption to the smooth flow of production material caused by managerial decisions (Hurley and Kadipasaoglu 1999: 1). Sometimes inappropriate performance measures encourage decisions that disrupt, rather than facilitate, the smooth flow of material through the factory. This can result in the *appearance* of a bottleneck. One example is running large batches to minimize the "cost" of setup. The result is material that cannot move before the batch is completed. Another example is expecting 100% utilization of labor and equipment. All protective capacity is eliminated, and any glitch disrupts the planned production output. These situations falsely indicate a bottleneck when the buildup of WIP is really due to managerial decisions.

Many of these popular fad approaches *seem* to make good business sense. They have contributed positively to the field of manufacturing, and many case studies verify their usefulness in specific applications. However, canned solutions tend to address a specific problem or set of problems. Each has an intended purpose and limited applicability. Applying these approaches one after another, searching for a *cure-all for manufacturing ills*, is what Shapiro refers to as "fad-surfing in the boardroom" (Shapiro 1995). The fad approach to manufacturing improvement leads to disappointment, cynicism, and inevitable resentment among managers and shop floor workers. It does not lead to holistic, long-term improvement of a factory.

## Cakes and Cookies

We often use the example of a small bakery (Peggy's Pantry) to illustrate the drawbacks and dangers of various "canned" solutions. Peggy's Pantry sells party cakes and giant cookies to the local grocery store, Favorite Foods. Our challenge is to determine the most profitable mix of cakes and cookies for the bakery. The owner of the bakery, Peggy, works 8 hours per day, five days a week. The fixed cost of keeping the bakery running is $1,500 per week. This includes the salaries paid to Peggy and to Elmo, the delivery boy. The bakery is fortunate because its customer, Favorite Foods, will accept any number of cakes or cookies that is less than or equal to a predefined maximum. For example, the grocery store will accept 50 cakes and 100 cookies but no more.

Peggy knows she has been losing money and would like to make the bakery more profitable, so she starts keeping detailed records. She plans to use this information to make better business decisions. Table 3.2a presents relevant production information about Peggy's Pantry.

| Product Information | Cakes | Cookies |
|---|---|---|
| Max. weekly sales (pieces) | 50 | 100 |
| Selling price ($) | 35 | 18 |
| Raw material cost ($) | 10 | 8 |

*Table 3.2a Bakery Product Information*

Recently Peggy read an article about cost analysis. This article suggested that factories should focus on making those products that are most profitable. This approach seems quite logical, so she decides to manage her bakery accordingly. First she must decide whether cakes or cookies are more profitable. Using the production information in Table 3.2a, it is easy to compute which product is the more profitable. For this simple comparison we define profit as the raw material cost subtracted from the selling price.

Profit per cake = \$35.00 − \$10.00 = **\$25.00**
Profit per cookie = \$18.00 − \$8.00 = **\$10.00**

From this analysis it is clear that *cakes* are far more profitable than cookies. Therefore, Peggy decides to bake all the cakes the customer will buy and then make cookies for the rest of the week.

We know that Favorite Foods will buy a maximum of 50 cakes, but there are several important questions to ask. For example, is it even feasible for Peggy's Pantry to make 50 cakes? If so, how long will it take to make them? How much time will remain for making cookies after the cakes are made? These questions can be answered by performing some simple calculations.

The total time available for making cakes and cookies is computed as shown below.

$$8\left(\frac{\text{hours}}{\text{day}}\right) \times 60\left(\frac{\text{minutes}}{\text{day}}\right) \times 5 \text{ days} = 2400 \text{ minutes}$$

Peggy knows how long each step in the cake-making process takes (Table 3.2b). The total processing time used at each station for making cakes can be found by multiplying the processing time per cake by the number of cakes, in this case 50. The time available for making cookies can be found by subtracting the time required for baking cakes from 2400 minutes. Table 3.3 shows the total processing time required to make 50 cakes and the remaining time available for making cookies at each processing station.

| Process | Cakes | Cookies |
|---------|-------|---------|
| Mixing | 20 | 10 |
| Baking | 45 | 10 |
| Cooling | 25 | 10 |
| Icing | 10 | 10 |

*Table 3.2b* Bakery Processing Time (Minutes)

| Process | Time Required to Make 50 Cakes | Time Available for Making Cookies |
|---------|-------------------------------|-----------------------------------|
| Mixing | 1000 | 1400 |
| Baking | 2250 | 150 |
| Cooling | 1250 | 1150 |
| Icing | 500 | 1900 |

*Table 3.3* Bakery Time Distribution

Since none of the total processing times in Table 3.3 is greater than 2400 minutes, Peggy concludes that making 50 cakes is feasible. However, after baking 50 cakes, the oven is only available for 150 additional minutes. This means that the oven can only bake cookies for 150 minutes total. How many cookies can be baked in 150 minutes? Again, the answer can be computed easily as shown below.

$$150 \text{ minutes} \div 10 \left( \frac{\text{minutes}}{\text{cookie}} \right) = 15 \text{ cookies}$$

So, the total delivery to the customer will be 50 cakes and 15 giant cookies. How much money will Peggy's Pantry make each week using this management approach? To find out, we simply add the profit from 50 cakes to the profit from 15 cookies and subtract the $1500 fixed expense.

($25 × 50) + ($10 × 15) − $1,500 = − **$100**

This is very disconcerting. According to the analysis of costs, Peggy's Pantry is destined to lose $100 each week!

Fortunately Peggy doesn't give up. She has noticed that no matter how hard she tries she can never bake the maximum quantity of both cakes and cookies. Something is *constraining* her from producing 50 cakes *and* 100 cookies all in the same week. She reads another article about factory management, and this one explains the importance of "bottleneck" operations when managing constrained systems. Could bottleneck analysis help her failing bakery? She decides to give it a try. First, however, she must determine which operation is the "bottleneck." Table 3.4 shows the amount of time required at each processing station to make 50 cakes *and* 100 cookies.

| Process | Time Required to Make 50 Cakes | Time Required to Make 100 Cookies | Total Time Required (Cakes and Cookies) |
|---------|-------------------------------|-----------------------------------|------------------------------------------|
| Mixing  | 1000 | 1000 | 2000 |
| Baking  | 2250 | 1000 | 3250 |
| Cooling | 1250 | 1000 | 2250 |
| Icing   | 500  | 1000 | 1500 |

*Table 3.4* Total Bakery Time for 50 Cakes and 100 Cookies

Baking 50 cakes and 100 cookies would require 3250 minutes of baking time. Since this is clearly greater than the 2400 minutes of available time, Peggy identifies the oven as the bottleneck. Then she defines a performance measure called "profit per minute of bottleneck time." Peggy wants to make the most profitable use of the bottleneck, so she uses her performance measure to determine which product should have priority.

Cakes: $25 / 45 minutes = **$0.56** profit per minute of bottleneck time.
Cookies: $10 / 10 minutes = **$1.00** profit per minute of bottleneck time.

So, based on her new performance measure, *cookies*, not cakes, should have priority. Peggy's new management plan is to produce as many giant cookies as the customer will buy (in this case 100) and to make cakes with the extra time.

Is it feasible for Peggy's Pantry to make 100 cookies? How long will it take to bake them? How much time will be available to make cakes after all the cookies are made? The answers to these questions can be found by performing the same simple calculations performed earlier.

Each giant cookie requires 10 minutes of baking time, so baking 100 cookies uses only 1000 minutes of oven time. This leaves 1400 minutes of oven time available for baking cakes. How many cakes can be baked in 1400 minutes? The answer can be easily computed as shown below.

$$1400 \text{ minutes} \div 45 \left(\frac{\text{minutes}}{\text{cake}}\right) = 31 \text{ cakes}$$

This means Peggy's Pantry can produce 100 giant cookies and 31 cakes in a single week. How much money will the bakery make if this management approach is taken?

($25 × 31) + ($10 × 100) − 1500 = **$275 profit**

By focusing on profitable use of the bottleneck operation, the bakery went from losing $100 to making $275 profit each week! Peggy is thrilled and gains faith in the bottleneck approach.

Peggy's Pantry operates this way for several weeks. Then the customer, Favorite Foods, asks Peggy to change the recipe of the giant cookies. The new type of cookie must cool twice as long on the cooling rack before the icing can be applied. The

| Process | Cakes | Cookies |
|---------|-------|---------|
| Mixing  | 20    | 10      |
| Baking  | 45    | 10      |
| Cooling | 25    | 20      |
| Icing   | 10    | 10      |

*Table 3.5 Revised Bakery Processing Times (Minutes)*

new processing times are shown in Table 3.5. Notice that only one entry, the cooling time for cookies, has been changed.

Peggy is not concerned about this minor change. After all, she is armed with a powerful analysis tool: bottleneck analysis. Table 3.6 is similar to Table 3.4 except the new cooling time for cookies has been used. It shows the total amount of time required at each processing station to bake 50 cakes *and* 100 cookies.

| Process | Time Required to Make 50 Cakes | Time Required to Make 100 Cookies | Total Time Required (Cakes and Cookies) |
|---|---|---|---|
| Mixing | 1000 | 1000 | 2000 |
| Baking | 2250 | 1000 | 3250 |
| Cooling | 1250 | 2000 | 3250 |
| Icing | 500 | 1000 | 1500 |

*Table 3.6 Revised Total Time for 50 Cakes and 100 Cookies*

Now Peggy notices that the time required at two processing stations, baking and cooling, exceeds the available 2400 minutes. In fact, both baking and cooling require 3250 minutes. Not knowing which one to choose, she decides arbitrarily to consider the cooling rack as the bottleneck. Again, she uses her new performance measure defined as profit per minute of bottleneck time.

Cakes: $25 / 25 minutes = **$1.00** profit per minute of bottleneck time.
Cookies: $10 / 20 minutes = **$0.50** profit per minute of bottleneck time.

This time cakes provide the most profit per minute of bottleneck time, so cake production is given priority. Peggy's Pantry should make as many cakes as the customer will buy, in this case, 50. Any time not used for making cakes should be used to make giant cookies. However, Peggy is puzzled because she realizes this policy is the same as the one suggested by cost analysis. She already knows this policy causes the bakery to lose $100 per week! Peggy wonders why the bottleneck analysis led her astray, and she decides that she must have chosen the wrong station as the bottleneck. Once again she computes her new performance measure, but this time she uses the oven as the bottleneck.

Cakes: $25 / 45 minutes = **$0.56** profit per minute of bottleneck time.
Cookies: $10 / 10 minutes = **$1.00** profit per minute of bottleneck time.

Peggy feels more comfortable with this analysis since it shows that cookies, rather than cakes, should have priority. The policy now suggested by the bottleneck analysis is to make 100 cookies and spend the remaining time making party

| Process | Time Required to Make 100 Cookies | Time Available for Making Cakes |
|---------|-----------------------------------|--------------------------------|
| Mixing | 1000 | 1400 |
| Baking | 1000 | 1400 |
| Cooling | 2000 | 400 |
| Icing | 1000 | 1400 |

*Table 3.7 Revised Bakery Time Allotment*

cakes. As before, the time available for making cakes can be found by subtracting the time required for making cookies from the total time available during the week (2400 minutes). Table 3.7 shows the total processing time required to make 100 cookies and the time available for making cakes at each processing station.

The cookies require 2000 minutes of time on the cooling rack, so only 400 minutes are available for cooling cakes. Since each cake must stay on the cooling rack for 25 minutes, the bakery can only cool 16 cakes. Nevertheless, because of Peggy's successful experience with the bottleneck approach, she is confident that this policy will generate a profit.

$$(\$25 \times 16) + (\$10 \times 100) - 1500 = -\$100$$

This policy is no better than the one suggested by cost analysis! Peggy now begins to fear she will lose her fledgling bakery. Then she remembers an important aspect of the bottleneck approach. The article said that if the constraint was too binding on the factory, the "bottleneck" should be "broken." With the help of her delivery boy, Elmo, she designs a special blower mechanism that will cool a party cake in 20 minutes flat instead of the 25 minutes now required. Table 3.8 shows the new processing times for the various production steps.

| Process | Cakes | Cookies |
|---------|-------|---------|
| Mixing | 20 | 10 |
| Baking | 45 | 10 |
| Cooling | 20 | 20 |
| Icing | 10 | 10 |

*Table 3.8 Improved Bakery Processing Times (Minutes)*

We leave it as an exercise for our readers to verify that the bottleneck approach still does not find a profitable production schedule regardless of whether the oven or the cooling rack is chosen as the bottleneck.

At this point, Peggy is tempted to tell Favorite Foods she cannot make the recipe change without first increasing her selling price. Before she does this, however, she remembers some wisdom she heard at a *Running Today's Factory* workshop: *a good solution applied to the wrong problem cannot lead to optimal performance.* She begins to wonder if perhaps she had been too hasty to apply the popular "bottle-neck theory," and she runs a few simple spreadsheet scenarios. Using some basic principles and her favorite spreadsheet program, Peggy finds the optimal solution in a few hundredths of a second. She *can* make the recipe change as requested by her customer and still generate a $155 profit each week. Furthermore, if she and Elmo can reduce the cake-cooling time to 20 minutes, the bakery's profit will increase to $210 per week, even using the new recipe!

Why didn't the bottleneck theory lead her to the right product mix? This is a case of the "good solution, wrong problem" paradox. The bottleneck approach does not always lead to optimal policies, particularly when there are multiple con-straints in the system. Applying canned solutions to situations where they aren't appropriate does not lead to competitive advantage. It is better to understand the basic principles of manufacturing and develop appropriate and effective solutions for the problems facing today's factory.

The bottleneck analysis *was* helpful in focusing attention on certain steps in the operation, namely baking and cooling. Improving a bottleneck can be an excellent way to improve a manufacturing operation; however, again we caution our readers that the approach is not a holistic improvement strategy.

## Robustness

If a solution is ideal and provably optimal but only applies to a narrowly defined situation, it is of little practical use to a manufacturing manager. On the other hand, if a solution is near optimal and can be applied with confidence to virtually any manufacturing situation, it is of great value. A robust solution is one that provides excellent results in a variety of situations with a wide range of input variables. A robust solution may not necessarily be optimal, but the user may be confident that the solution is sound and near optimal in virtually any situation. Robustness must be considered when evaluating a manufacturing solution. The bottleneck approach was not a robust solution to the problems of Peggy's Pantry.

Consider an illustration from the field of civil engineering: the placement of reinforcement bar imbedded in reinforced concrete. Precise placement of reinforce-ment bar is critical to the integrity of the concrete. For example, if a concrete slab is to be used as a bridge to span a garden stream, the "re-bar" should be placed quite low in the slab. In this way, the steel re-bar can bear the tensile stresses in the slab. In fact, a rule of thumb valid for all *similar* situations is to place the re-bar on the side of the slab that is away from the load. A problem arises, however, when this rule of thumb is applied to a *different* situation. Consider now a retaining wall that is to be poured in the same garden. If the re-bar is placed on the side opposite the load (the outside face), the retaining wall will soon crumble! There is nothing inherently wrong with the rule; it was just applied to the wrong situation. The rule is *robust* when applied to spans such as bridge decks, floors, or roofs of buildings, but the rule does not apply universally to all concrete structures.

Here is an example of a *robust* rule for running today's factory: *it is always beneficial to reduce variability throughout the production process.* Reducing variability results in shorter cycle time, shorter lead time, lower inventory, lower production costs, higher quality, greater flexibility, higher throughput, and more profit. Much of this book is devoted to explaining this rule and providing proven strategies for reducing variability in the factory.

## Job Shop Scheduling

Another "canned solution" that is often applied in factories is the first-come, first-served (FCFS) scheduling algorithm. FCFS has the advantage of seeming fair, and it is usually applied without much further analysis. (There are several other approaches to job shop scheduling, and each has its own performance characteristics. For example, EDD (earliest due date first) and SPT (shortest processing time first) are common scheduling algorithms that perform well in many circumstances.) Consider a local job shop, *Marino's Machine Shop*, that supplies a variety of industrial customers on a make-to-order basis. Over the past few days, a good customer, *Needit-now*, ordered five items from Marino's. Let's assume that each of these "jobs" must be processed through the same two machines (Machine 1 and then Machine 2). Table 3.9 shows the processing time for each job at each of the two machines.

| Job | Machine 1 | Machine 2 |
|-----|-----------|-----------|
| A | 7 | 5 |
| B | 6 | 6 |
| C | 7 | 4 |
| D | 5 | 2 |
| E | 5 | 10 |

*Table 3.9 Processing Times for Each Job (Hours)*

Marino's is very busy this particular week and does not start on any of the jobs for Needit-now. The orders are entered into Marino's computer system, and they are scheduled to run in FCFS sequence the following week. Then, Needit-now calls in a panic needing *all* items as soon as possible. Marino's Machine Shop stops work on all other jobs and schedules the Needit-now jobs immediately. Will the FCFS scheduling method result in the shortest completion time for all five jobs?

In operations research, the time required to produce a given number of jobs is called *makespan*. In this case, minimizing the time Needit-now will have to wait is equivalent to minimizing makespan. What sequence of jobs minimizes makespan? Before discussing the answer, we will provide some background information about this entire class of problems.

The optimal solution to this type of problem was first published by Johnson in 1954, and there is some interesting folklore surrounding his solution. The legend begins with Johnson having a discussion with his wife about housework. Evidently there were several loads of laundry that needed to be washed and ironed, but Johnson had planned a day of golf. His wife had other plans. Johnson was instructed that he had to do the laundry before he could play golf. Being a scholar, Johnson decided to whip up a quick algorithm to sequence the washing and ironing in order to minimize the total time required (makespan). By following this optimal sequence, he could get to the golf course at the earliest possible moment. Surprisingly, the solution proved to be less than obvious, and he ended up spending the whole day working on his algorithm. He never made it to the golf course, and (guess what) he never did the laundry. He did, however, develop a provably optimal sequence that is still taught and talked about in the operations research community 55 years later!

We should point out that the story surrounding Johnson's solution is only folklore. He was actually solving a real-world problem posed to him by a colleague at the RAND Corporation. It was known as the bookbinding problem. However, more than one researcher has observed that when a golf game hangs in the balance, minimizing makespan seems to be *much* more important than it is in bookbinding (Dudek, Panwalker, and Smith 1992)!

Now, back to Marino's Machine Shop. Obviously, the total processing time re-

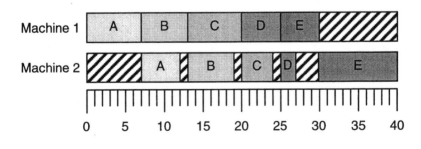

**Figure 3.1a** *Time Line for Processing All Jobs in First-Come, First-Served Order*

quired at each machine is not affected by the sequence of the jobs. Waiting time, however, is affected dramatically by the sequence of jobs. Let's look at a time line of these five jobs processed in the sequence in which they were ordered by Needit-now (FCFS) (Figure 3.1a).

The total time to run all five jobs in the FCFS sequence is 40 hours. Notice the idle time when Machine 2 is waiting for a job to process. Also notice the idle time when Machine 1 has finished all jobs, but Machine 2 is not finished yet. Minimizing makespan is equivalent to minimizing the idle time on Machine 1 and Machine 2.

Johnson's rule tells us to look at the time required to process each job at each machine. If the shortest processing time occurs on Machine 1, then that job is scheduled before the others. If the shortest processing time is on Machine 2, then that job is scheduled last. The remaining jobs are analyzed in the same manner. The job having the next shortest processing time is identified. Again, if this time is on Machine 1, that job is scheduled in the first available position. If the time is on Machine 2, the job is scheduled in the last available position. This procedure repeats until all jobs are scheduled. We should point out that if a job is scheduled first on Machine 1, it is also scheduled first on Machine 2. In fact, the sequence that is determined for Machine 1 is the identical sequence that must exist on Machine 2. In other words, jobs do not preempt each other.

In our example of Marino's Machine Shop, the shortest processing time is 2 hours (Job D, Machine 2). Since this time occurs on Machine 2, Job D is scheduled last. The next shortest time is 4 hours (Job C, Machine 2). Job C is scheduled immediately before Job D. The next shortest processing time is 5 hours (Job A, Machine 2; and Job E, Machine 1). Job A is scheduled immediately before Job C, and Job E is scheduled first. By default, the remaining Job B is scheduled in the only available slot, immediately after Job E.

Scheduling according to Johnson's algorithm reduces the time to run all five

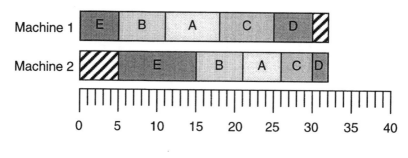

Hours of Processing Time

***Figure 3.1b*** *Time Line for Processing All Jobs Using Johnson's Rule*

jobs to 32 hours (Figure 3.1b). This is 20% shorter than the time needed by the FCFS schedule! Johnson's approach minimizes the makespan and the customer waiting time, and it also highlights the impact that waiting time has on makespan. Although Johnson's algorithm is provably optimal, there are not many practical situations where it can be applied. Consequently, it cannot be considered robust. The extension of Johnson's rule to three-machine or to n-machine systems becomes extremely complicated mathematically. Johnson himself had little success developing other algorithms for job shop scheduling—but not for lack of trying. Along with scores of other operations researchers, Johnson spent years trying to solve this class of problems (Dudek, Panwalkar, and Smith 1992: 11).

Job shop scheduling falls into a class of problems known as *hard*. In operations research, the term "hard" goes beyond merely being difficult. "Hard" connotes being difficult to program numerically. "NP" is an abbreviation for "nonpolynomial." A problem is said to be NP-hard if the time to find a solution grows as an exponential or nonpolynomial function of the number of input variables.

Consider a job shop scheduling problem only slightly more complicated than the Marino's Machine Shop example. This time, let's say the 5 jobs will be processed through 3 machines instead of 2. We will also say that the jobs can be processed through the machines in any order. For each machine there are 5! (or 120) different possible orderings. If jobs can be processed on the machines in any order, it follows that there are $(5!)^3$ or 1.7 million possible schedules. Now let's say that the 5 jobs need to be processed through 6 machines. Again, for each machine there are 5! (or 120) different possible orderings. If the jobs can be processed on the machines in any order, it follows that there are $(5!)^6$ or 3 trillion possible schedules! Even with the latest computing capability, finding the optimal solution to fairly simple job shop scheduling problems is impossible or impractical at best. Now consider a more reasonably sized but still simple job shop scheduling problem: 10 jobs and 10 machines. There are $(10!)^{10}$ or $4 \times 10^{65}$ potential schedules. This problem was not solved until 1988 (Carlier and Pinson 1988).

Most factory scheduling problems involve hundreds of jobs and dozens of operations and are indeed NP-hard. Finding optimal solutions to these real-world problems is *quite literally impossible*. Hopp and Spearman have eloquently stated that "a computer with as many bits as there are protons in the universe, running at the speed of light, for the age of the universe, would not have enough time to solve some of these problems" (1996: 472).

## Summary

What's so tough about running a factory? The obvious answer is that a factory has problems. However, even if these manufacturing problems didn't exist, the job of running the factory would still be tough. Lack of agreement about what the factory is supposed to do and how it is supposed to do it, diverse and conflicting demands, outdated manufacturing dogma, short-term performance measures, and "canned solutions" that lead us astray all add to the difficulty of running a factory. The environment is changing rapidly, and running a factory is a lot more complicated than most people realize. Optimal solutions for even surprisingly "simple" factory problems "simply" cannot be found, even with the latest computer programs. Fortunately we do not always need to find the *optimal* solution. An *excellent* solution—meaning one that helps us meet our manufacturing and business goals—is usually good enough.

The research community has been working to solve operations management problems for many years. Until very recently, however, most research has concentrated on solving hypothetical problems that are of little practical use. A noted researcher in operations management makes the following observation: "Once the war was over, academics rapidly discovered that it is easier (and safer) to solve problems that had no relevance to reality" (Woolsey 1993: 41). While abstract research is valid scientifically and has advanced the frontier of operations research, it has not made the job of running a factory much easier.

There are no quick fixes for the problems faced by today's factory. Many of the popular manufacturing fads work well in certain situations. They are often substantiated with sensational case studies, but scientific rationale is usually not given. For this reason, they are difficult to challenge or refute. Nevertheless, the American manufacturing community *as a whole* has not realized the promised benefits. Why? We believe the problem lies not with the approaches *per se* or even their implementation. The problem is that *canned solutions are not holistic solutions*.

Fortunately, *there are holistic approaches to running today's factory* that can lead to dramatic overall improvement in performance. One of these is based in philosophy and involves the way we *perceive* the factory. It requires us to think about manufacturing in a completely different way. This philosophical approach emphasizes a clear understanding of goals and a common strategy. It has a central theme that can be used to help make manufacturing decisions. This approach, known as lean manufacturing, leads to robust and near optimal solutions.

The other holistic approach to running today's factory is a scientific approach. There is an entire *science of manufacturing* that is emerging from the fields of operations research and queuing theory. This scientific approach provides a framework for understanding how manufacturing systems behave and how different policies and decisions affect overall performance. This science of manufacturing is known as factory physics, and we discuss it extensively in Chapter 5. The remainder of this book explores the philosophical and scientific approaches to running today's factory.

*Chapter 4*

# Lean Manufacturing Overview

## Introduction

The term "lean" has been used extensively in the manufacturing community for more than twenty years. It has been commonly associated with low inventory levels and reduced staffing. Factory employees are especially sensitive to this term since "lean" has been used sometimes as a euphemism for layoffs. In 1988 an entirely different meaning was attached to the term "lean." In a landmark article published in the *Sloan Management Review*, John Krafcik introduced "lean" to describe a philosophy of production (1988: 45). This production philosophy has since proved to be as important and as distinct as the American System or mass production.

Krafcik, who was a member of the MIT International Motor Vehicle Program (IMVP), trained in California at NUMMI (a Toyota–General Motors joint venture) and spent a great deal of time at Toyota in Japan. He initiated a study known as the World Assembly Plant Survey and gathered data from ninety plants in seventeen countries. It is considered the most comprehensive international survey ever conducted in any industry (Womack, Jones, and Roos 1990: 5-6, 75-76).

The book, *The Machine That Changed the World,* summarizes the five-year five-million-dollar study of the international automobile industry. The title is a play on words. In one respect, the "machine" that changed the world is the automobile. By providing inexpensive, personal transportation, the automobile completely transformed societies and their economies on a global scale. In another sense, the "machine" is not a machine at all but rather the way automobiles are produced. In the first half of the twentieth century, mass production in various stages of evolution revolutionized industry and industrialized nations. In the latter half of this century, a *different* but equally significant approach to production developed in the Japanese automotive industry. This approach is so radically different from traditional mass production that it is changing fundamentally the way goods are produced worldwide. The IMVP researchers recognized and called attention to this new manufacturing paradigm, and they named it "lean production." In effect, automobile production has changed the world *twice* this century: first by the development of mass production, and second by the development of lean production.

Many people are familiar with lean manufacturing and its tantalizingly powerful techniques. Impressive case studies have been published in excellent books and countless journal articles. These numerous case studies document the dramatic success some companies are having as a result of implementing lean manufacturing. Awareness of these lean success stories leads to an appreciation of the impressive results, but it can also lead to misconceptions about how those results were achieved. Case studies alone cannot provide holistic understanding of lean philosophy.

This chapter presents a concise definition of lean manufacturing and a clear explanation of its central theme and overall objectives. We discuss the origins of lean manufacturing, including how and why it developed. We describe what it looks like on the shop floor and its substantial and concrete benefits to the factory. We strip away much of the mystery that surrounds lean manufacturing by conveying its fundamental objectives and addressing common misconceptions. This chapter concludes with a brief discussion of culture and the role culture plays in today's factory.

## Lean Manufacturing

Quite simply, lean manufacturing is a production philosophy, a fundamentally different way of *thinking* about manufacturing. It is a different way of conceptualizing the entire production stream from raw material to finished goods and from product design to customer service. It is a philosophy of manufacturing that seeks to minimize unnecessary time, materials, and effort in the production process.

Lean is unlike previous production philosophies such as craft production, the American System, or mass production (Chapter 2). The most obvious differences are how suppliers supply the factory, how material "flows" through the stream of production, and how the factory delivers to the customer. For example, lean manufacturing depends on more frequent deliveries of smaller quantities from fewer suppliers. Lean factories produce in smaller batches and alternate production items more often. Products are sent to customers more frequently and in smaller lots. These are the most obvious differences; however, the essence of lean is much deeper. *Lean manufacturing is actually based on an entirely different value system.*

The values underlying lean manufacturing are distinct from the values underlying more traditional forms of manufacturing. In lean manufacturing, what is perceived as beneficial, conscientious, prudent, safe, and economical is often considerably different from what is advocated in other styles of manufacturing. What is viewed as harmful, wasteful, dangerous, costly, and imprudent may also be different. Manufacturing values influence behavior: behaviors that are consistent with the value system are encouraged, and behaviors that are contrary to the value system are discouraged. The values themselves arise from different underlying assumptions about manufacturing and how a factory should operate. These underlying assumptions are often subconscious. They are so deep in our belief system that they are accepted as ultimate truths, and they are rarely questioned or challenged.

Consider a value that is at the very heart of mass production: *economies of scale*. According to the mass production paradigm, large production lots are considered *economical* because setup time is amortized over many parts. This maximizes the capacity utilization of the equipment and increases the "efficiency" of the production department. Accepting delivery of a large quantity of parts minimizes the receiving effort, and moving large lots around the factory minimizes material handling effort. All of these functional areas benefit from the "economies of scale" associated with large lots.

According to the lean production paradigm, *large production lots are costly rather than economical*. This is because they tie up production resources for a long time and disrupt the smooth and continuous movement of material through the factory. This hinders flexibility and necessitates an excessive amount of work-in-process inventory (WIP). The excess WIP requires storage space and support operations such as stacking, moving, staging, counting, and re-packing, and the material is subject to damage, obsolescence, and loss. Since there is a time lapse between when the items are made and when they are used, defects may not be discovered until long after the items are produced. This makes it difficult to find the root cause. Also, if any items are defective, a large amount of WIP implies a large number of defects, and, therefore, a large amount of rework and scrap. These costs easily overshadow any "efficiency gains" attributed to large lot production.

The way the factory is measured or evaluated is also very different in lean manufacturing. Traditional measures such as equipment utilization and standard labor variance are irrelevant (Chapter 12). In the lean manufacturing paradigm, different goals are pursued. Enlightened measures such as cycle time, worker participation, and sales per employee-hour track the progress of the factory as it strives to meet the manufacturing challenge. In lean manufacturing, even the definition of a "problem" is quite different, and, consequently, different solutions are developed. *Lean manufacturing truly is a different way of thinking about manufacturing.*

It is difficult for people to think about familiar situations from a new and different point of view. This is a human tendency that is generally true of all people anywhere in the world. As adults we have our customary ways of perceiving what happens and making judgements about the events we experience. We no longer need to be intensely aware of every experience because, based on our years of "data," we can make generalizations about what takes place and quickly categorize events as normal/abnormal, beneficial/harmful, good/bad, and so on. Cognitively speaking, as human beings we have perceptual templates that we use to interpret and understand the world around us. These templates develop from infancy through adulthood based on our experiences and cultural values. This is how we make sense of our social and physical environment and know how to respond. This is a highly efficient way for humans to exist in a complex environment, but it makes it very difficult for us to change the way we evaluate, judge, and react to familiar situations.

For example, each of us has a set of criteria by which we evaluate people based on their clothing. Are they savvy, successful, appropriate, organized, interesting, approachable? The dress codes of our society change constantly, but our criteria as individuals change much more slowly, if at all. Even though we realize our fashion template may not be 100% current, it is very difficult not to react to others according to our own set of rules. The definition of "dress for success" changes all the time. For example, many companies now promote a "business casual" dress code, and people wear clothing to work that once was appropriate only for the annual picnic. Changing the norm does not happen instantly. People have to get accustomed to it, and some people find it very hard to react in a different way.

Similarly, the perceptual templates we use for understanding *manufacturing* situations are formed early in our careers and often early in life. They are constructed subconsciously over time through our experiences and through our exposure to production situations. We observe how products are made, why problems occur and how they are solved, how the factory is managed, what outcomes are rewarded or discouraged, what behaviors are successful, and what constitutes a *well-running* factory. Some of us have been taught formally about production management through training or academic courses or by managers who mentored us. Others have developed on-the-job intuition about how the factory should run. No matter how our perceptual manufacturing templates were developed, those of us who work in manufacturing have them, and we use them to evaluate and assess situations in the factory. Since lean manufacturing is a *different way of thinking*, running today's factory requires those of us who are accustomed or acculturated to mass production philosophies to relinquish our existing paradigms and open our minds to new possibilities and new interpretations.

*Lean manufacturing provides the highest quality at the lowest cost in the shortest amount of time.* "Highest quality" is made possible by ensuring that what is produced is correct rather than inspecting, sorting, fixing, or monitoring quality. "Lowest cost" is made possible by eliminating wastefulness such as scrap, unnecessary effort, additional floor space, waiting time, and excess work-in-process inventory (WIP). "Shortest lead time" results from making only the required amount with perfect quality, delivering the product to the proper place at the proper time, and being flexible and responsive to changing customer demand.

What does lean manufacturing look like? This style of production is called *lean* because it uses "less of everything" (Womack, Jones, and Roos 1990: 13). The MIT study showed that lean production requires *half as much engineering effort, product development, and design time.* Products are designed with manufacturing in mind and with the participation of factory personnel and suppliers. Consequently, less time and effort are spent designing and redesigning, and the products are easier to manufacture. In a lean environment it is unacceptable to manufacture products that are difficult to make. Such products require either a complicated production process or extensive redesign; both alternatives are expensive and time consuming. Lean factories need only *half the human effort for the same production output.* This is because workers spend less time performing operations that do not contribute to the value of the product. Also, *half the factory floor space* is needed. Factories that become lean can double output without enlarging their facilities or adding workers. Lean factories have *one-tenth as much work-in-process inventory (WIP)* because production material flows continuously from raw stock to the customer's dock. Lean factories receive production material from *one-tenth as many suppliers.* Suppliers are not selected solely because they bid the lowest price or promise the earliest delivery date. Instead, a long-term partnership is established with a few suppliers based on mutual consideration of each other's business needs (Schonberger 1986: 155-59). Lean manufacturers are able to produce their goods in *one-tenth the total processing time.* Again, this reflects the emphasis on continuously flowing material and the elimination of wasted effort.

In short, products are designed for easy production, and they are manufactured in a timely manner. Components of certified quality are always available. They are delivered in small quantities on a regular basis from a few cooperative suppliers. The factory floor is uncluttered and well organized. Only the necessary amount of work-in-process accumulates between processes. The processes themselves are grouped by product family rather than by function. Operators work at a steady pace performing tasks that add value to the product. The organizational structure of the company is based on product families rather than functional departments such as "Sales" or "Welding." Problems are corrected as soon as they are recognized, and the production process is continually improved. A lean factory has a very different appearance from a traditional factory and can be recognized immediately by even a casual visitor.

## Benefits of Lean Manufacturing

Lean manufacturing provides an extraordinary opportunity to improve quality, customer service, and profitability. Unfortunately, incorporating lean concepts into an existing manufacturing operation can be an arduous task that strains the very fabric of the organization. While researchers freely acknowledge that a lean transformation is not particularly easy, the difficulty associated with the endeavor is often understated. "The evidence indicates that the transferability of these methods to environments outside of Japan has been *sometimes problematic*" (Daniel and Reitsperger 1996: 96). Jeffrey Liker, Director of the University of Michigan's Value Chain Analysis Program and Japan Technology Management Program, is somewhat less sanguine in his assessment of the transformation process. "The adoption and implementation of lean manufacturing, however, is complex, slow, incremental, and unpredictable, and the process varies dramatically from case to case" (1997: 498). Because of this difficulty, many manufacturing managers believe a lean transformation would be *too costly*, and the benefits would not justify the effort. Joseph Day, the CEO and President of Freudenburg-NOK, has a very different opinion:

> "Even if they can get past the concept's incongruity with their basic mass manufacturing tenets, they fear lean implementation requires too much—too much time, too much change in the corporate culture, and too much money. In the heat of daily business, some executives see the conversion to lean manufacturing as a luxury they simply can't afford. As a CEO who has wrestled with those same issues, I found that converting to lean systems was the one thing our company couldn't afford *not* to do" (Day 1994).

What makes lean manufacturing worth the effort? Why would a company go through a difficult transformation just to adopt a lean philosophy? The answers to these questions are self-evident for those companies that are making the transition. A recent manufacturing survey reveals several of the competitive advantages companies are realizing as they transform their operations to lean manufacturing (Struebing 1997). Table 4.1 presents several advantages of lean manufacturing and the percentage of companies realizing each particular advantage as they strive to become lean.

| Selected Competitive Advantages | Companies Realizing Advantage |
|---|---|
| Reduced customer lead time | 63% |
| Steady or reduced pricing | 63% |
| Increase in market share | 61% |
| Reduced time to launch new products | 39% |
| Increased product diversity | 24% |

*Table 4.1* Selected Advantages Realized by Lean Companies

Another important advantage that companies realize as they become lean is a marked increase in productivity. This means that for a given level of resource utilization, such as equipment, people, time, space, and investment, lean manufacturing results in much higher productive output. The efficiency gains associated with lean manufacturing are of particular interest to many companies and labor associations. Often fewer workers are needed to maintain a given level of throughput. The issue that is of great concern to many shop floor workers and managers alike is what to do with the displaced workers. The same manufacturing survey mentioned above also addresses this issue. Table 4.2 lists the various actions taken by companies after they realized a productivity increase and the percentage of companies surveyed that chose each particular action.

| Action Taken After Productivity Increases | Companies Taking Action |
|---|---|
| Increased production and sales | 69% |
| Guaranteed no layoffs due to increased efficiency in productivity | 92% |
| Reduced overtime | 73% |
| Reduced the number of temporary employees | 33% |
| Reduced the number of employees through attrition | 50% |
| Transferred displaced employees to improvement teams | 38% |
| Used displaced employees for product development | 13% |

*Table 4.2* Actions Taken After Productivity Increases

Generally speaking, manufacturing companies that are becoming lean are realizing tremendous benefits. These benefits bring important competitive advantages such as reduced lead time, lower manufacturing cost, lower selling price, increased market share, and increased profit. Now let's see exactly how several companies are benefiting as a result of their lean manufacturing efforts.

The Critikon Company is a division of Johnson & Johnson located in Southington, Connecticut. The management declared a "war on waste" in 1991 and embarked on an aggressive factory transformation. During the following four years, Critikon made some very dramatic improvements (Robson 1994):

- 94% improvement in supplier quality
- 50% increase in productivity per employee
- 53% reduction in scrap
- 50% reduction in setup time
- 60% reduction in cycle time
- 30 % reduction in inventory cost
- 50% reduction in WIP
- 156% increase in training hours per employee
- 30% reduction in required floor space per unit
- 50% reduction in warehouse space.

Freudenberg-NOK established its now famous GROWTTH Program in 1992. GROWTTH is an acronym for Getting Rid Of Waste Through Team Harmony. Between 1992 and 1996, Freudenberg-NOK conducted 2,500 *kaizen* "events" in fifteen plants involving 90% of their workforce. Sales tripled to $600 million, and profits soared to unprecedented levels. During the first two years of the transformation, the CEO and President, Joseph Day, devoted 35% of his work week exclusively to *kaizen* implementation (Liker, 1997: 179 and 188). These are some of the accomplishments:

- 35% reduction in setup time
- 47% reduction in lead time
- 58% reduction in cycle time
- 43% reduction in down time
- 44% improvement in productivity
- 33% reduction in WIP
- 32% reduction in floor space
- 45% reduction in scrap
- 82% reduction in distance traveled.

A recent article in *The New York Times* focused on the benefits resulting from lean implementation in the aerospace industry (Pollack 1999). At the Boeing factory in Long Beach, California, "production of the C-17 military transport has been doubled in the last three years without raising the head count." The Northrop Grumman Corporation reduced one mechanic's job on the B-2 bomber from 8.4 hours to 1.6 hours by kitting the supplies needed for the task and making the supplies available at the worksite.

The B. F. Goodrich factory in Chula Vista, California, consolidated production activities and reorganized the shop floor by product rather than by function. Production activities that were performed in five buildings spread over a mile were consolidated into two buildings. These changes resulted in greatly reduced production cycle times and far less distance traveled on the shop floor. For example, the distance traveled by one component of the Boeing 717 fan cowl has been reduced from 17,000 feet to 4,300 feet, and its production cycle time has been reduced from 43 days to 7 days. "Owing in part to lean production, operating profit margin for Goodrich's aerostructures division has risen to 15.6 percent in 1998 from 9.9 percent in 1995" (Pollack 1999).

These improvements are typical of a factory undergoing a lean transformation. The benefits of lean manufacturing can be quite dramatic, and this explains why so many companies are undertaking the difficult transition. In an increasingly competitive environment, lean manufacturing can provide advantages that are not easily duplicated by rival firms.

So, it is really not a question of whether or not to transform today's factory; it is a question of *how* to transform today's factory. Unfortunately, there is no easy answer. The literature concerning the transference of lean principles and practices to existing factories provides no clear instruction or consensus. Most articles and books on this subject are descriptive or anecdotal. While such accounts verify methods and document benefits, they offer little assistance to the manufacturing manager who is developing a transformation plan.

Only recently have researchers begun to apply the scientific method to the subject of factory transformation. In Chapter 13 we develop a roadmap for transforming today's factory using the best of this research and relying heavily on our own factory experience. It is important to remember that success is not achieved by instituting tactics or techniques; it begins by adopting a nontraditional way of *thinking* about manufacturing.

## The Origin of Lean Manufacturing

Lean manufacturing is not new, and, in fact, some aspects of it are nearly 200 years old. Product-oriented layout, work standardization, continual improvement, built-in quality, and short cycle time are common lean practices, and *all* of them can be traced to earlier periods of American manufacturing.

For example, one important aspect of lean manufacturing is *training for flexibility*. This results in multiskilled workers who are capable of job rotation. The benefits of multiskilled workers and job rotation were enumerated in an 1854 British evaluation of the American System (Wilson 1998: 84). Two major benefits noted were: 1) having multiskilled workers provides flexibility for the factory, and

2) job rotation relieves monotony. Interestingly, these are the same two reasons we give for establishing flexible work cells in factories today!

Another important aspect of lean production is the unilateral disdain of overproduction. In fact, overproduction is considered by lean manufacturing proponents to be the worst waste of all. There is a good reason: overproduction leads to excessive WIP and a plethora of related problems! Two hundred years ago, the American System view of overproduction was strikingly similar. In the American System "production schedules seem to have been driven largely by sales orders, [so] there was little latitude for overproduction" (Wilson 1998: 85).

Consider the Highland Park Ford plant, which began producing the Model T in 1910 (Chapter 2). This plant was an early example of *product-oriented layout*. In other words, the arrangement of equipment and workstations followed the sequential order in which the processes occurred. In a product-oriented layout, material flows unimpeded from one process to the next. While this one principle does not define lean manufacturing, it is completely consistent and compatible with lean principles. It is ironic that Ford abandoned this idea for process villages when the Rouge Complex was constructed, it was a 100-year step backwards for American manufacturing.

One of the most important concepts in lean manufacturing is *short cycle time*. Henry Ford recognized the importance of cycle time and continuous flow. He even boasted that iron ore from the mine could be converted into a finished automobile in 81 hours (Hopp and Spearman 1996: 28). This shows a remarkable commitment to short cycle time. Lean manufacturing didn't actually develop at Ford Motor Company, but it certainly could have. Many lean concepts were applied at different times in the company's early history. The similarities between lean manufacturing and the early Ford system "are so impressive that Henry Ford may honestly be considered a pioneer of JIT systems" (Wilson 1996: 30).

Another example from American manufacturing is Training Within Industry (TWI), a U.S. government program established during World War II. Drastic changes, whether social, scientific, or industrial, are often preceded by a catalyst, some crisis that is so important, urgent, or perhaps even life threatening that our minds become receptive to a new way of doing things. For the United States, such a catalyst was our involvement in World War II which led to an urgent need for increased industrial output and the creation of TWI.

TWI was established in order to boost U.S. industrial productivity and output significantly enough to win the war. The program taught supervisors to solve their own manufacturing problems, thereby improving efficiency. Some key elements of the TWI job methods were standardized work, elimination of unnecessary details (waste), efficient work sequences, pre-positioning of tools at the point of use, gravity-feed hoppers and drop-delivery chutes, and the use of standard work methods until better ways are developed (continual improvement). All of these elements are completely consistent with lean manufacturing philosophy (War Manpower Commission 1945: 37-38).

There were 1.7 million American supervisors from more than 16,000 plants trained and certified in the TWI program. These supervisors, in turn, trained over 10 million factory workers. Almost 90% of the participating companies achieved at least a 25% improvement in production and labor efficiency. In 1945 the TWI program in the United States was deactivated by the War Manpower Commission (War Manpower Commission 1945: 92, 126).

In 1945 there was a different crisis in Japan. Japan's industrial activity had been reduced to less than one-tenth of its pre-war level, and the economic base had been devastated. If the economy was to recover, the country needed to rebuild its industrial infrastructure. As part of the assistance provided by the Occupation forces, TWI was introduced. By 1951 the Allies were relying on Japanese textiles, metals, and auto industries for supplies during the Korean War, and the TWI effort was intensified. The following year there were over one million certified Japanese supervisors. In 1990 the program was still active with nineteen training organizations licensed by the Japanese Ministry of Labor (Robinson 1991: 21).

# The Toyota Production System

Lean manufacturing is not, as some researchers suggest, a generic term for the Toyota Production System (TPS). However, Toyota has made great contributions to the field of manufacturing management, and no treatise on modern manufacturing principles would be complete without discussing TPS. It is interesting that Toyota, a company that has become the paragon of lean manufacturing, incorporated many elements of TWI and practices from the American System. Toyota was, however, the first company to incorporate lean principles as a coherent and clearly articulated system of production. TPS continues to shine as an example of lean manufacturing in practice. Understanding the development of TPS is complicated by the mix of historical fact and legend. We present one version here.

Two individuals are prominent in the development of the Toyota Production System: Eiji Toyoda and Taiichi Ohno. In 1940 at the age of 17, Taiichi Ohno was a supervisor at Toyota Automatic Loom Works. He learned the principles of *jidoka* (quality-at-the-source) during his tenure there. In 1947 he moved to the Toyota Motor Company (TMC). At the end of World War II, TMC was the largest automotive producer in Japan. Compared to Western automotive companies, however, Toyota's production volume was extremely low. By 1950 Toyota had produced fewer than 2700 automobiles in the entire history of the company. That is less than what the Ford Rouge plant was producing in a single shift! Furthermore, Toyota's tiny production volume included many types of vehicles. The few automotive plants in Japan in 1950 had to supply extremely diverse transportation needs, such as cars, delivery vans, heavy trucks, light trucks, ambulances, limousines, and fire trucks. The challenge to the Japanese automotive producers and to Toyota in particular was to manufacture this diverse product mix effectively and efficiently.

In 1950 Ohno and others at Toyota estimated that American autoworkers were nine times more productive than their Japanese counterparts. Toyota management was reluctant to accept that the disparity was due to the workers themselves because Japanese textile workers were among the most productive in the world. Instead, Toyota concluded that the difference must be in the *system of production*. This led to Eiji Toyoda's historic pilgrimage to the Ford Rouge complex in Dearborn, Michigan. His objective was to learn the basis for American success and efficiency in automotive production, and to evaluate the feasibility of Ford's mass production system succeeding in Japan (Nicholas 1998: 13).

The American automotive market was virtually unlimited, and each assembly plant specialized in its own specific product family. For example, in 1950 the Ford Rouge plant was pumping out 7000 similar vehicles each day. This contrasted sharply with Toyota, which was producing many different vehicles in small volume. Toyota did not have the resources or the market to support many plants, and the product mix was too eclectic to justify dedicated plants.

The Rouge complex that Toyoda observed did not have a product-oriented layout like the earlier Highland Park facility had. That organizational approach had been abandoned for process villages with each functional area having its own "plant." Each process village produced in advance and in excess of what the downstream process actually required, so each process village accumulated its own stockpile. The assembly village, where the automobiles were assembled, drew from those stockpiles as needed. One result was that the assembly plant could produce a somewhat wider variety of vehicles since all necessary components were available in the many stockpiles. Another result was that *production material spent several weeks waiting to be used*. It is important to point out that production material stayed in the plant for many weeks even though the time required to transform material into a car was a matter of minutes. This meant that the time between purchasing raw materials and being paid by a customer was needlessly long.

For three months, Toyoda studied Ford's methods at the Rouge facility. He concluded that the Ford mass production system was impractical for Toyota. Multiple plants specializing in specific models could not be supported by the post-war Japanese economy. Even if Toyota had wanted to build a dedicated plant to accommodate each product line, it would not have been possible; financial resources were too scarce. Furthermore, the process village concept that had been adopted by Ford was completely unworkable in the post-war Japanese economy; any capital invested had to be converted to revenue as soon as possible. Accumulation areas and stockpiles of inventory cost far too much in terms of time and money to be a viable option for Toyota.

So, why didn't Toyoda take the Ford production system from the Rouge plant back to Toyota? There was too much material, floor space, time, and investment capital tied up for too long. Instead, the Toyota executives developed a completely *different way of thinking about manufacturing*. From 1950 to 1975, an alternative to the Ford mass production system matured in the rapidly recovering Japanese economy. Toyota led the way and developed a clearly articulated production system based on a *different philosophy of manufacturing*.

## Cycle Time

Toyoda concluded that Ford's methods would not work in Japan for a variety of reasons. One major reason was *cycle time*. (Cycle time is the length of time required for production material to flow through the factory from raw material to finished product. Cycle time is also known as sojourn time, flow time, throughput time, and dock-to-dock time. We use the term "cycle time" in order to be consistent with the operations research literature.) Ford's cycle time was long, not because so much work was required to transform the material into an automobile, but because so much time was *wasted* while material waited at various stages throughout the production process.

Generally speaking, *processing time* only accounts for a small fraction of total cycle time. By far the largest constituent of cycle time is *waiting* time. This can be illustrated by a simple game of golf. According to a club maker for Senior PGA Tour professionals, during a typical swing the clubface is in contact with the ball for approximately 0.02 seconds. During this brief instant, energy is imparted to the ball causing it to move (hopefully) toward the hole. A typical round of golf might take longer than four hours to play, but the ball is in contact with the clubface for less than 1.5 seconds total! The rest of the round is spent waiting: waiting for the ball to land, waiting for other members of the foursome to tee off, waiting to walk or drive the cart to the ball, waiting to find the ball, waiting for the group ahead to leave the green, and many other waiting delays throughout the game. The length of time required to play the game is analogous to cycle time, and the time during which the ball is in contact with the clubface is analogous to processing time.

Now, let's imagine our club maker could develop a special club that reduces contact time with the ball by 50%! Would the total time required to play a round of golf be shorter? A typical round of golf would still take more than four hours even though the total contact time was reduced to less than 0.75 seconds.

As this analogy implies, waiting time is also the largest component of cycle time in a factory. This wasted time is spent waiting in queues, waiting for tools, waiting for people, waiting for machines, waiting for material handling resources, waiting for inspection or rework, waiting for the other items in the lot to be finished, and many other waiting delays. The golf club illustration also highlights the folly of "improvement" activities that focus on decreasing the processing time at individual workstations. That approach completely ignores the largest component of cycle time, which is waiting time.

Consider a stamping facility. How much time does the die spend in contact with sheet metal to form a part? In a machining operation, how much time is the cutting tool actually cutting the workpiece? In both cases, the contact time is a miniscule portion of the overall cycle time; yet, we usually focus our improvement efforts on activities like hits per minute or spindle speed. Ironically many improvement efforts ignore the activities that consume most of the production time.

Why should we use our engineering resources to shave tenths of a second off processing time if the material is simply going to wait for days or weeks to be inspected, moved, counted, and stored? The purpose of nearly every element of lean manufacturing is the reduction of cycle time. This leads us to the core strategy of lean manufacturing: *to achieve the shortest possible cycle time by streamlining the flow of production material!*

## Value-Added Activity

Production time and activity can be classified as either adding value or not adding value to the product. *Value-added activity* is effort that has value in the eyes of the customer. It can be activity that transforms material into a product, or it can be a service for which the customer will pay. Stamping, painting, anodizing, and assembly are examples of value-added activity. Surprisingly most production effort has absolutely no value from the customer's perspective and is classified as *non-value-added activity*. Staging, stacking, counting, reworking, moving, and storing are activities for which a customer is not willing to pay. Another type of non-value-added work activity is *incidental work*, and although it doesn't add value to the product, it is a necessary part of the process. Examples of such activities are processing purchase orders, loading material into a machine, reaching for a tool, packing, and billing.

## Production Time Line

The time line in Figure 4.1 represents production time in an automotive production facility. Let's follow a single piece—a body panel—through the entire production process. First, a raw coil of steel comes from the supplier and waits to be processed. Then, a blanking press cuts the steel into many blanks. Our particular blank waits with thousands of other blanks for further processing. Many days or weeks later, stamping dies form our blank into a shaped body panel ready for welding. But the body panel doesn't go directly to welding; it waits. And waits. Perhaps the panel deteriorates or gets damaged and must be reworked. Finally, the body panel reaches the welding department where it is welded onto the body of an automobile. Then the car body waits. The body may deteriorate or may become damaged. If this happens the body is repaired, after which it waits. Then the body is painted, and it waits. Perhaps the body was painted a color that is not needed immediately, so it waits. Possibly the finish is not acceptable, and the body must be repainted. In the assembly area, the final assembly takes place. Then the new car waits. After the value-added processing is completed, the automobile is shipped to the dealer inventory lot where it waits. Perhaps it does not have the options the customer wants, so it waits until finally (maybe) a customer arrives who wants the option package on that particular automobile and likes the paint color and the model.

As can be seen from Figure 4.1, only a fraction of the total cycle time is spent transforming raw material into an automobile. Most of the time is simply wasted. People are busy moving, storing, stacking, and repairing the damage that takes place along the way. It is evident why this scenario does not work in an economy where cash is scarce and investments must be converted into revenue as quickly as possible.

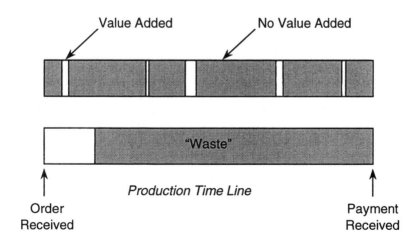

*Figure 4.1* Production Time Line in a Typical Factory

Figure 4.2a illustrates how manufacturing effort is spent in a typical factory. The pie chart represents total manufacturing effort. Some of that effort is value-added work for which the customer will pay. (Note that the value-added slice has been enlarged greatly to make it visible. Value-added activity is usually an infinitesimal portion of the total manufacturing effort!) Other effort is incidental and must be done even though it doesn't add value to the product. By far the largest portion of manufacturing effort is wasted effort.

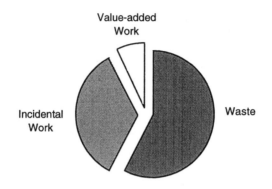

**Figure 4.2a** *Total Manufacturing Effort in a Typical Factory*

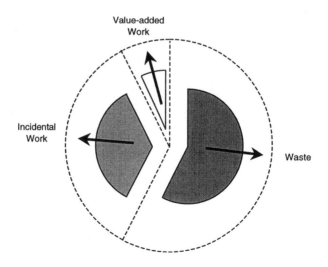

**Figure 4.2b** *Increasing Value-Added Work by Increasing Total Effort*

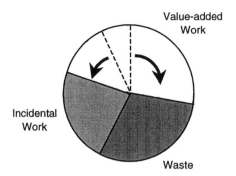

**Figure 4.2c** *Increasing Value-Added Work by Decreasing Incidental Work and Eliminating Waste*

"How can our factory increase the amount of value-added work?" The traditional answer is, "Do more of everything." Hire more people. Add a shift. Pay workers for overtime. Make the workers work harder or faster. Buy another machine. Run the machinery longer. As we can see in Figure 4.2b, this definitely increases the amount of value-added work, but it also increases the incidental work and the wasted effort (time, space, material, and money). The lean manufacturing answer is, "Increase the value-added work by reducing incidental work and by eliminating wasted effort" (Figure 4.2c). Time and effort that were previously spent on non-value-added activities are redirected to value-added activities. The factory operation is the same size. The workers don't work harder or longer. There is no added equipment. A shift is not added, nor are more workers hired. This is the strategy for running a successful factory today.

## Respect for Humans

One central philosophy for running today's factory is a profound respect for the human beings working there. In fact, the first English language article ever written about the Toyota Production System uses the term "respect-for-human system." The article refers repeatedly to *employee responsibility*, often in conjunction with the term *authority* in the same sentence. A clear assertion is that assigning responsibility without commensurate authority is demoralizing and disrespectful (Sugimori, Kusunoki, Cho, and Uchikawa 1977). The principles and practices for running today's factory involve not only efficient material flow and increased productivity but also involve increased respect for the workers and reliance on their human potential.

It is unfortunate that in many factories terms such as "employee involvement" have been so overused in rhetoric and underused in actual practice that they have become meaningless at best and a sham at worst. A true respect for the people involved is absolutely crucial in any manufacturing environment promoting continual improvement. People need to feel that they are partners in the factory's success.

Furthermore, each worker represents a wealth of expertise and experience and is an important asset to any manufacturing organization. People can be seen as a sustainable competitive advantage. Systems, equipment, automation, and software can be duplicated by the competition. People, on the other hand, are unique in their knowledge, and their collective contributions represent a significant advantage to those companies who recognize it and capitalize on it.

## Meeting Diverse Customer Demand

Meeting the diverse needs of customers is not a new challenge. In Chapter 2 we traced the different methods manufacturers have used to meet this challenge throughout American manufacturing history. The lean approach, however, is distinct from that taken by mass producers. Let us turn our attention to how American automotive producers meet this challenge today.

Many American factories still produce in anticipation of what the customer will want. (It is important to recognize that the "customer" does not necessarily need to be the end user; the customer can also be the next process in the production sequence.) The central theme of mass production has become "make in anticipation of what the customer will want while maximizing economies of scale." In fact, the entire production stream from raw material to finished goods is managed this way in a mass production enterprise.

When the system behaves in a suboptimal manner (and it always will!), the problems are often blamed on inaccurate forecasts. Larger, faster, and more expensive computer systems are often purchased to remedy the situation and supposedly make the operation more profitable. This spiral can continue endlessly with MRP, MRP-II, ERP, and related computer solutions being installed and upgraded on increasingly sophisticated computer systems. Considerable time, effort, and money are wasted, and interdepartmental relationships suffer due to the inevitable finger pointing that takes place when the new system fails to achieve the promised results.

No matter how sophisticated the forecast-based system, the fact remains: *producing in anticipation of what the customer wants results in an excess of unneeded material and a scarcity of material that is actually required by the customer or next process.* Even though the production system becomes bloated with inventory, customers are constantly disappointed because they cannot get what they want. Furthermore, waiting for the factory to make what the customer *does* want takes an inordinate amount of time because the factory is busy making what the customer doesn't want.

The lean approach is to take the manufacturing challenge literally: *make what the customer wants when the customer wants it and sell it at a price the customer is willing to pay*. This means making a wide variety of goods from a single production facility, as efficiently as possible with the proper amount of work-in-process. The overarching theme is the highest quality at the lowest cost in the shortest amount of time.

## Common Misconceptions

Lean manufacturing is currently receiving considerable attention. It is addressed in many publications from the daily newspapers to academic research journals. Several universities (the Universities of Kentucky, Tennessee, and Michigan, and the Georgia Institute of Technology, to name a few) offer lean manufacturing courses and seminars through their schools of business or engineering. In 1997 James Womack founded the Lean Enterprise Institute in Brookline, Massachusetts. Its mission is to promote *lean thinking* throughout entire corporations and to provide a forum where lean practitioners can discuss ideas and issues. The American Supplier Institute (ASI) offers a complete lean manufacturing curriculum including technical workshops and holistic executive seminars to help managers transform their factories successfully. The Society of Manufacturing Engineers (SME) holds several lean manufacturing seminars each year. One area of concentration for APICS certification is "just-in-time." In 1998 more than five conferences on lean manufacturing were held in the greater Detroit area. Some were instructional, and others offered case studies from companies where lean transformations are under way.

Ironically, the growing popularity of lean manufacturing is accompanied by widespread misconceptions. This is not because the principles themselves are complicated or difficult to understand. The misconceptions arise from limited exposure, unclear definitions, and the application of isolated techniques. An incomplete understanding of the principles can lead to skepticism and rejection of the entire philosophy. Some people who have a passing acquaintance with lean manufacturing fear that *becoming lean* will cause the factory to incur greater costs. We often hear the comment that "lean manufacturing is great, if you can afford it." To help clarify the lean manufacturing philosophy, we have chosen three of the most common misconceptions for closer examination.

One misconception is that lean manufacturing was a fad that has now been superceded by more contemporary fads. Some who are out peddling their own brand of "solution" characterize lean as outdated or *passé* because the term was introduced more than ten years ago. This has resulted in a rash of new books, articles, and new terminology such as "agile manufacturing" and "after lean." These attempts to appear trendy are confusing and add little to our understanding of how to run today's factory. As we indicated earlier in this chapter, many practices encouraged by lean philosophy originated between 50 and 200 years ago! Lean manufacturing is not new. It is not a quick fix or a redecoration project. It is a maturing philosophy of production that provides clear principles for running today's factory.

Another misconception is that lean manufacturing is simply a collection of various Japanese manufacturing techniques. Although the philosophy was developed in Japan, lean is not necessarily "Japanese." As we have seen in the cases of the American System, the Highland Park Ford plant, and TWI during World War II, many of the techniques have been used in the United States. Consider the support for just-in-time (JIT) strategies in Japan and in the U.S. It probably won't surprise our readers that the majority of Japanese managers favors the JIT principles. What is surprising is that an even greater percentage of U.S. managers also favors the JIT approach. While there is little evidence that JIT is being used widely in the U.S., there is strong evidence that American companies *intend* to use JIT. In fact, JIT as an *intended* strategy is now more prevalent in the U.S. than in Japan (Daniel and Reitsperger 1996: 109). Ironically, most Japanese factories do not use lean manufacturing, and most American manufacturers know more about it than their Japanese counterparts know (Yamada 1998).

The last misconception we address is that lean manufacturing is an assortment of tools and techniques from which factory managers can select or "cherry pick." Using a technique should never be a goal for the factory. Techniques are only the means by which a factory can achieve its goals. While visiting factories, we are often shown the visible evidence of a lean technique such as a work cell, a *kanban* card system, a mistake-proofing device, a quick-change fixture, or an area of the factory that has been cleaned and organized. This is disturbing if the changes being imposed onto the factory are not part of a holistic improvement effort. Using a lean tool by itself may result in some outward semblance of lean manufacturing, but it cannot result in a lean transformation without being part of a comprehensive, long-term strategy.

One example of a misused tool is *visual factory*. The term "visual factory" often means workplace organization. This may include tool kits conveniently placed where they are used, or painted areas of floor space reserved for specific purposes. These changes are apparent immediately and "look lean," but visual factory is actually a much deeper concept. A visual factory is self-aware, and any abnormalities are immediately and automatically communicated throughout the plant. Workplace organization by itself does not lead to lean manufacturing. On the other hand, lean factories do tend to have excellent workplace organization and a marked improvement in shop floor appearance. This is a result, however, not the cause, of the factory being lean.

# Cultural Considerations

It is widely believed that production methods are not cross-culturally applicable; therefore, lean manufacturing can be successful only in Japan where the philosophy developed. We feel this is an important issue worthy of careful consideration. Cultural values are based on religion, tradition, philosophy, experience, and the underlying assumptions that are basic to that specific culture. These values, in turn, give rise to culturally acceptable patterns of behavior and interaction. Culture influences greatly the way people work, interact with one another, and interpret the world around them.

Let's consider briefly the cultural values that are prevalent in Japan. Japanese social life has been influenced strongly by the Buddhist, Confucian, and Shinto religions and by a strong tradition of cooperation and mutual obligation and respect. Some values from Japanese culture are particularly relevant to daily work life. These include harmonious relationships (*wa*), compassion, respect for others, seeing others as an extension of oneself, clarity, calm, continual improvement (reincarnation), practicality, and absorption in the task at hand. Interestingly, these cultural values are intrinsically compatible with lean manufacturing.

One example of traditional obligation is the cooperative farming of terraced rice fields. The success or failure of an entire community's rice crop depends on intensive teamwork, communal responsibility, and each individual's diligence in maintaining the irrigation system. One researcher attributes the cooperative work ethic in Japanese factories to the collective interdependence of villagers farming the rice terraces. "It is this rice culture that we now see in the Japanese factory" (Morgan 1986: 115). Whether or not one accepts Morgan's interpretation, the example does illustrate the pervasive values of cooperation and mutual obligation in Japanese culture.

The traditional relationship (*rentai*) of mutual obligation between managers and workers is of primary importance in all business relationships including those in the factory. In this relationship, each group has specific responsibilities to the other (Kim and Takeda 1996). Management obligations include allocating daily job responsibilities, evaluating performance, providing potential lifetime employment, treating all workers equally and as part of a family. Worker responsibilities include following orders without mistakes, cooperating with co-workers, generating ideas to improve the organization, and respecting upper levels of management. In short, everyone maintains the best possible relationships.

Clearly, culture influences attitudes, work ethic, and interpersonal relationships in the factory, but an important question remains. Is cultural compatibility enough to explain the success of the lean approach to manufacturing in Japan? A recent study of Japanese manufacturing companies reveals that neither cultural values nor just-in-time (JIT) techniques alone is sufficient to produce the tremendous performance improvements associated with lean manufacturing. Instead, the effectiveness seems to come when propitious cultural values are blended with the practices. Companies that are rich in "Japanese" cultural values do not achieve world class performance levels unless the elements of lean manufacturing are also introduced. More importantly, the application of lean techniques without a corresponding culture change "will create 'islands of JIT,' and will hinder the total implementation process" (Kim and Takeda 1996).

So, successful implementation of lean manufacturing does not require the culture to be "Japanese," and lean transformations can proceed successfully outside Japan. However, as lean strategies are adopted in the boardroom and as lean practices are applied to the shop floor, certain cultural values must be introduced and encouraged. Incorporating a value system that is consistent with lean manufacturing philosophies is an important factor for success. This value system must reinforce appropriate goals, patterns of behavior, and work habits on the shop floor. People will not change the way they work or their underlying assumptions about manufacturing if they continue to be appraised and promoted according to the previously existing value system.

Let's contrast the values advocated by the mass production culture with those embraced in a lean manufacturing culture. Notice how the mass production values revolve around individuals and individual efficiencies.

- More is better.
- Faster is better.
- Overproduction is good.
- If a process is going well, keep doing it.
- Don't stop the production line.
- Take stopgap measures, but don't stop to fix the problem.
- The front line workers are responsible for production output.

Notice how the lean production values emphasize cooperation and total system efficiency.

- Make the next person's job easier.
- Make what is needed when it is needed.
- Overproduction is evil.
- If a process is going well, make it even better.
- Never pass on a defect.
- Stop production to fix the problem.
- The managers are responsible for enabling production workers to do their jobs effectively.

It is difficult to change a person's value system. Values are so central to an individual's identity that they may be totally subconscious, and any attempt to modify them will be resisted strongly. This is predictable, understandable, and a perfectly natural part of being a human. As shown above, certain aspects of lean manufacturing are in opposition to traditional mass production values. This can cause great resistance if the value system underlying people's work behavior is attacked directly. People clinging to their old ways of doing things and defending their old values can hinder or even halt a lean transformation. Changing someone's core value system takes a long time.

It is far easier and much faster to modify behavior. Behavior modifications can be encouraged and supported by a system of rewards, recognition, and reprimands. So behavior changes can be made and reinforced even before the underlying values have changed. (Our discussion of TWI emphasized that a number of key elements of lean manufacturing originated in the United States. When that program was exported to Japan, the transfer was successful even though some American cultural values, such as a democratic management style, were intrinsic to the approach.) Eventually an individual's attitudes and value system may change as the new behavior patterns become the norm. Meanwhile, the desired behaviors are in place and the factory and its people are benefiting from the lean transformation.

## Summary

Mass production brought together many existing principles and practices with a fundamental goal: *to maximize efficiency by exploiting economies of scale.* Likewise, lean manufacturing has a fundamental goal: *to achieve the shortest possible cycle time by streamlining the flow of production material throughout the value stream.* Lean philosophy seeks to maximize the efficiency of the entire value stream. The central theme of lean manufacturing is to do more with less: less time, less inventory, less space, and fewer resources. Core values are respect for people and appreciation of human creativity. The lean goals and philosophies are generally applicable in today's business environment, which is why lean thinking is so powerful and effective in highly diverse industrial situations.

The growing popularity of lean manufacturing has led to numerous misconceptions about it. One misconception is that it is simply a passing fad; but lean manufacturing is not new, it is not a passing fad, and it is still relevant today. Another misconception is that "lean" is simply the "Japanese" approach to manufacturing. While lean manufacturing did develop in Japan, it is not intrinsically Japanese. Many of the concepts were used earlier in American manufacturing. A third common misconception is that lean manufacturing is an assortment of independent tools and techniques from which factory managers can "cherry pick." This overlooks the fact that lean manufacturing is a holistic approach to running today's factory. Lean principles can be used to develop sound manufacturing strategies, and these strategies can provide long-term competitive advantages.

This chapter introduced the concept of cycle time, which is the time required for material to flow through the factory. By far the largest component of cycle time is waiting time. Manufacturing effort can be viewed as either adding value or not adding value to the product. By redirecting effort from non-value-added activities to value-added activities, the productive output of the factory can be increased without investing in additional resources or expanding the current operation. Wasted effort extends cycle time far beyond the actual processing time for any given product. Reduction of cycle time is particularly important when a company needs to convert investments into revenue quickly.

A society or a company may have a value system that is highly compatible with the underlying philosophies of lean manufacturing, and an appropriate cultural environment certainly contributes to success. Fortunately, a factory can create and encourage a suitable culture through its system of recognition and rewards. Culture, however, is not enough to produce the tremendous benefits of lean. A propitious environment must be blended with appropriate measures and goals, long-term corporate plans, and sound principles and practices.

Companies that have initiated a lean transformation are reaping tremendous rewards. Typical benefits include lower production costs, higher quality, shorter cycle times, shorter customer lead times, greater flexibility, better customer service, and higher profit. More important than short-term gains, lean thinking can be used to develop an overall, long-term business strategy. This strategy can be the key to running today's factory successfully (Chapter 6).

# *Chapter 5*
# **Factory Physics Basics**

## Introduction

There are no generic solutions for the challenges and dilemmas facing today's factory. Good manufacturing decisions result from careful observation, rigorous analysis, and well-developed intuition. Unfortunately, the onslaught of management buzzwords confuses the issues. Quick fixes and clever slogans promise dramatic improvements but do not offer coherent guidelines for implementation. Many current management fads have no scientific basis; they rely solely on anecdotes and elaborate rhetoric to "prove" their value and effectiveness. Blind faith is tough to defend in the boardroom, and many managers have become cynical with good reason.

Fortunately, lean thinking provides a sound philosophical basis for establishing appropriate policies and making manufacturing decisions. As Chapter 4 emphasizes, lean manufacturing is much more than a collection of good ideas; it is a systemic philosophy of manufacturing with a permeating central theme. This central theme is the reduction of cycle time by streamlining the flow of production material throughout the value stream.

Interestingly, lean ideology is completely consistent with solid, proven scientific principles from operations research and queuing theory. This chapter presents the underlying scientific principles that explain objectively *why* lean thinking is such an effective approach to manufacturing management. The principles and practices presented in this book provide both philosophical and scientific bases for running today's factory.

Within the general research field of production and operations management, there is an emerging specialty known as *factory physics*. Factory physics is sometimes called the "science of lean manufacturing" although it evolved quite independently. Based on direct observation, factory physics first develops *descriptive models* that characterize factory operations. Then *prescriptive models* are developed that can guide future manufacturing decisions. The intent is to use a rigorous, analytical approach to consciously build sound intuition about manufacturing problems.

The term "factory physics" was introduced to the industrial and academic communities by Wallace Hopp and Mark Spearman (1996). Their approach has been called "the most radical and nontraditional model of production and operations management" (Leschke 1998). We agree! *Factory physics is a radical departure* in the following sense: most operations researchers make certain basic assumptions and then derive mathematical models based on those assumptions regardless of whether or not the models represent a realistic manufacturing situation. Such models are often tautological with the "proof" contained in the statement of the problem. This academic approach emphasizes the derivation of mathematical equations rather than the solution of practical problems. Factory physics, on the other hand, is more pragmatic. It reveals the underlying principles that govern observable factory behavior, and provides analytical tools for making sound manufacturing decisions.

There is a wide disparity between the problems being solved in the research community and the problems facing today's factory manager. In a review of flowshop-sequencing research, Dudek, Panwalkar, and Smith (1992: 9) admit that they "have *never been approached by anyone* claiming to have a need for solving a problem having the characteristics assumed by most flowshop researchers." To better appreciate their point, let us list the typical assumptions made by most flowshop researchers.

- Jobs or work orders are processed sequentially through $m$ stations.
- There is one machine at each station.
- Machines are available continuously.
- A job is processed at one station at a time without preemption until completion.
- Only one job is processed at a station at a time.

Usually flowshop research investigates the efficiency of scheduling. It is concerned with static, deterministic sequencing problems in which $n$ jobs are processed through $m$ stations; the processing time at each station is fixed (constant) and known. However, most factory managers we know deal with a tremendous amount of variability. They schedule varying numbers of jobs through varying numbers of different machines, and the actual processing times are uncertain and highly variable!

Solutions to fixed, theoretical situations offer little insight into a real factory where *variability is the only constant.* Dudek, Panwalkar, and Smith comment about the proliferation of research that ensued after Johnson (1954) published his elegant job shop algorithm (Chapter 3). "At this time, it appears that one research paper (that by Johnson) set a wave of research in motion that devoured scores of person-years of research time on an intractable problem of little practical consequence" (Dudek, Panwalkar, and Smith 1992: 10).

Although the operations research literature is replete with articles on almost every aspect of production management, we have noticed that the salient points are rarely applied in industry. This is largely because many of these articles are *unavailable.* They are often published in journals not read routinely by factory managers. Another reason the conclusions are not applied is *irrelevance.* The elegant models and algorithms offered by the research community are derived for hypothetical situations that rarely occur in actual factories. A third reason is *specificity.* Most articles focus on solving highly specific problems rather than developing a *general science of manufacturing* that can be used to build intuition and guide decision making. In fact, there is not yet a generally accepted science of manufacturing although several researchers are beginning to identify some basic manufacturing "laws." While this field is admittedly incipient, as a scientific approach it provides useful insight into the behavior of manufacturing systems. It builds intuition and confidence, and it provides a rationale for making manufacturing management decisions.

In this chapter we present the basics of factory physics. We illustrate this scientific approach to factory management with several examples. In each case, rigorous analysis from the perspective of factory physics leads to the same conclusion we reach using lean thinking. Consequently, factory physics can be used to further substantiate and corroborate decisions made through lean thinking. Factory physics is much more powerful than a philosophy can be; it provides scientific direction even when lean thinking is ambiguous.

For example, lean thinking emphasizes the "elimination of waste," but in some cases, strategically placed waste in the value stream actually helps the factory reach its competitive goals! The advantages can outweigh the disadvantages. Factory physics demonstrates that the factory must compensate for variability by having extra inventory (waste), extra capacity (waste), or longer lead times (waste). One company might see its ability to respond quickly to changing customer demand as an important competitive advantage. To achieve this flexibility the company might choose to invest in extra inventory (waste) or extra capacity (waste). In essence, one company's waste is another company's competitive advantage.

We are very familiar with an aerospace components manufacturer that considers shorter lead time to be a competitive advantage. Because of this, the company's manufacturing strategy is to reduce cycle time wherever possible. In one particular area, cycle time was reduced greatly (70%) by *increasing* the material handling effort. So, by adding extra handling (waste) to the process they were able to eliminate waiting time (waste) and achieve their corporate goal of shorter lead time.

## Basic Factory Dynamics

Introductory physics courses usually concentrate on kinematics and mechanics, which involve the relationships among force, motion, and energy. The concepts are conveyed in lectures on classical or *Newtonian* physics and by hands-on laboratory experiments in which students can "discover" the basic laws of motion and build intuition about the physical world.

Unfortunately, laboratory sections in operations management courses are rare, and most factory managers learn about factory dynamics through direct observation "on the job." The basic laws of factory physics are difficult to observe in an actual factory. They are obscured by many uncontrollable factors that influence factory behavior. For this reason, a *simulated* factory is often useful to help managers and shop floor workers understand the basics of factory dynamics. This can be done with computer simulation, but we usually prefer a hands-on approach. By studying the simulated factory the basic laws become apparent, and valuable insight is gained about the behavior of an actual factory.

Before discussing how factory physics concepts can be applied to a factory situation, we will define some important terms: *cycle time, throughput,* and *work-in-process inventory* (WIP). Cycle time is defined as the amount of time material spends in the system. It is measured in time-units such as hours, minutes, or seconds. (See Chapter 4 for a more detailed discussion of cycle time.) Throughput is the rate at which material is processed through the production system. It is measured in product-units per time-unit such as pieces per hour. Another key parameter in factory physics is work-in-process inventory. In this context WIP should be measured in product-units such as pieces. Now we can perform a *mental exercise* that will illustrate clearly how these three factory parameters are related.

First, consider a simple four-station production line in which the processing time at each station is exactly 1 minute. Let's assume the factory produces in batches of 10 pieces at a time. How long will it take to complete all 10 pieces? Each piece requires 1 minute of processing time at the first station, so a total of 10 minutes is required to process the batch of 10 pieces at that workstation; 10 minutes are required to process the batch at the second station. In fact, if we ignore the transportation time, it would take exactly 40 minutes to process the entire batch of 10 pieces through all four workstations. Each piece spends 40 minutes in the system, so the cycle time is clearly 40 minutes. The throughput is 10 pieces every 40 minutes or 0.25 pieces per minute (15 pieces per hour).

What will happen to cycle time and throughput if the factory produces in batches of 4? Each piece still requires 1 minute of processing time at the first station, so a total of 4 minutes is required to process all four pieces through the first workstation. Similarly, 4 minutes are required to process the batch at the second station, and 4 minutes are required to process the batch at the third station. Again, if we ignore the transportation time, it would take exactly 16 minutes to process the entire batch of 4 pieces through all four workstations. Each piece spends 16 minutes in the system, so the cycle time is clearly 16 minutes. The throughput is 4 pieces per 16 minutes or 0.25 pieces per hour (15 pieces per hour). This exercise can be performed with any batch size, but we have tabulated the results for all batch sizes between 1 and 10 (Table 5.1).

| Batch Size (WIP) | Cycle Time (minutes) | Throughput (pieces per minute) | Throughput (pieces per hour) |
|---|---|---|---|
| 1 | 4 | 0.25 | 15 |
| 2 | 8 | 0.25 | 15 |
| 3 | 12 | 0.25 | 15 |
| 4 | 16 | 0.25 | 15 |
| 5 | 20 | 0.25 | 15 |
| 6 | 24 | 0.25 | 15 |
| 7 | 28 | 0.25 | 15 |
| 8 | 32 | 0.25 | 15 |
| 9 | 36 | 0.25 | 15 |
| 10 | 40 | 0.25 | 15 |

*Table 5.1* Batch Production Performance

Since material is transferred from station to station in batches, this style of production control is known as *batch and queue* or *batch and push* production. The performance of batch and queue production is known as *worst case performance* because, mathematically speaking, it is the worst possible way to process material through a factory.

In batch and queue production, the batch size has a pronounced effect on cycle time but absolutely no effect on throughput! This casts doubt on the anachronistic dogma we hear from shop floor workers and managers who insist that it is faster or more efficient to process in large batches. Another observation we can make about throughput based on this table is that throughput is always equal to WIP divided by cycle time. This "law" of factory physics is known as Little's Law in honor of John D. Little who published its mathematical proof in 1961. Little's Law is always true, even in the presence of random variability, and it can be applied either to a single processing station or to an entire routing. Little's Law is exhibited in Table 5.1.

Little's Law:
$$\text{Throughput} = \frac{\text{WIP}}{\text{Cycle Time}}$$

Sadly, some factories perform worse than "worst case." This occurs when the factory practices what we call *batch and store* production. In batch and store production, the material is not sent directly to the next process where it waits in queue for the station resources. Instead, it is produced in batch and moved to an off-line storage location where it may wait indefinitely. Manufacturers have euphemisms for this waiting material such as *fail safe* or *safety stock*.

We can gain even more insight into factory behavior by changing the production control policy and repeating the exercise. Under our new policy items are moved immediately from one workstation to the next as soon as they are ready. This style of production control is known as *one-piece flow, single-piece flow,* or *continuous flow* production. The performance exhibited by single-piece flow production is known as *best case performance*. Mathematically speaking, this is the best and most efficient way possible to process material through a factory. In fact, from a factory physics perspective the benefits promised by *cellular manufacturing* (Chapter 8) are largely due to its accommodation of single-piece flow manufacturing.

We should point out that arranging equipment or workstations into a "cell" does not guarantee best case performance. Recently we toured a factory in southwestern Michigan and observed a "manufacturing cell" that exhibited extreme worst case performance. The equipment had been arranged in a "U" shape, but the production material was moved from station to station in batches of 200 pieces rather than as individual pieces! Hopp and Spearman relate a similar experience with a "work cell" that produced in batches of approximately 10,000 pieces. "Thus, although the cell had been designed to reduce WIP and cycle time, the actual result was the closest we had seen to the worst case performance" (1996: 289).

Returning to our mental exercise, we will begin by setting WIP = 1, and, again, we start the clock as soon as the first item is introduced to the system. The first item is processed at the first station, which takes exactly 1 minute. Then it is passed immediately to the second station for processing. After another minute it is ready to move to Station 3. At the end of exactly three minutes it moves to the fourth station for its final operation. The total time this particular piece remains in the system is exactly 4 minutes, so the throughput is 1 piece per 4 minutes or 0.25 pieces per minute (15 pieces per hour).

What would happen if WIP = 2? Again, we start the clock as soon as an item is introduced to the system. This item is processed at the first station, which takes exactly 1 minute. Then it is passed immediately to the second station for processing. At the instant the first piece begins processing at Station 2, another piece begins processing at Station 1. These two pieces follow each other throughout the four-station line. It is easy to see that each piece remains in the system for exactly 4 minutes. When the first piece is completed, another piece is introduced at the first station. In this way the system contains two pieces of WIP at all times. Such a system produces exactly 2 pieces every 4 minutes or 0.50 pieces per minute (30 pieces per hour).

An interesting nuance occurs when the WIP level reaches 5. Let's assume there are already 4 pieces in the system, one at each station, and each of these pieces is ready to be processed. When the fifth piece is introduced at Station 1, it cannot be processed because the first station is processing the fourth workpiece. The fifth piece must wait until the fourth piece proceeds to Station 2. The fifth workpiece remains in the system for exactly 5 minutes: 1 minute waiting in queue at Station 1, and 4 minutes of processing time. Each time one piece is completed, another piece is introduced at the first station, and the newly introduced piece waits exactly one minute before *it* begins processing. In this example, the system has 5 pieces of WIP at all times and produces exactly 5 pieces every 5 minutes or 1.0 pieces per minute (60 pieces per hour). We performed this experiment for all WIP values between 1 and 10, and the results are shown in Table 5.2.

Table 5.2 indicates that increasing WIP does not result in a corresponding increase in cycle time until the system is full. There is a WIP threshold (in this case 4 pieces). If WIP is increased beyond this threshold, production material stacks up as it waits to be processed. This causes cycle time to increase, although not as

| Work-in-Process (WIP) | Cycle Time (minutes) | Throughput (pieces per minute) | Throughput (pieces per hour) |
|---|---|---|---|
| 1 | 4 | 0.25 | 15 |
| 2 | 4 | 0.50 | 30 |
| 3 | 4 | 0.75 | 45 |
| 4 | 4 | 1.00 | 60 |
| 5 | 5 | 1.00 | 60 |
| 6 | 6 | 1.00 | 60 |
| 7 | 7 | 1.00 | 60 |
| 8 | 8 | 1.00 | 60 |
| 9 | 9 | 1.00 | 60 |
| 10 | 10 | 1.00 | 60 |

*Table 5.2* Single-Piece Flow Performance

dramatically as in the batch production case. It is interesting to note that *reducing WIP below the threshold results in a corresponding decrease in throughput.* This casts doubt on the *mantra* of some manufacturing "gurus" who claim that WIP should be reduced to zero, or that zero inventory should be the goal! It is clear that increasing WIP improves throughput, but only to a point. (In this case, throughput increases until WIP = 4 pieces.) Once this WIP threshold has been reached, further increases in WIP do not improve throughput. As expected, Little's Law continues to hold.

The data from Tables 5.1 and 5.2 are presented as Figures 5.1a and 5.1b. Again, batch and queue production corresponds to worst case performance, and single-piece flow production corresponds to best case performance.

Typical factories perform between the two extremes of worst case and best case performance, although, as noted earlier, it is possible for factories to perform worse than worst case. Let's see what will happen if a typical factory makes an arbitrary policy to lower WIP. Figure 5.2a shows clearly that such a decrease in WIP would cause a corresponding decrease in throughput. In fact, we have often observed this phenomenon in actual factories. The plant manager responds to admonitions from the board of directors to "decrease inventories," and the result is lower throughput, lower revenues, worse customer service, and lower profit!

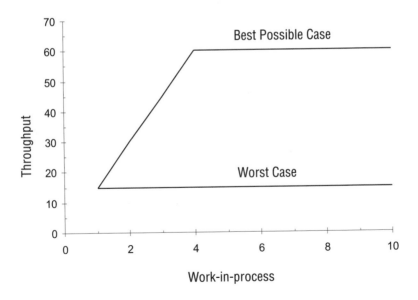

***Figure 5.1a*** *Throughput as a Function of WIP*

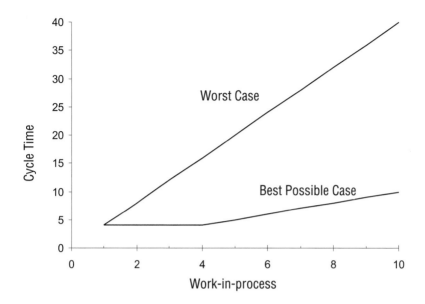

**Figure 5.1b** *Cycle Time as a Function of WIP*

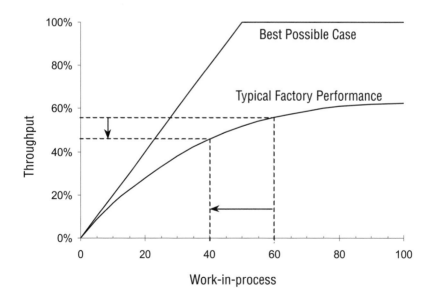

**Figure 5.2a** *Throughput as a Function of WIP*

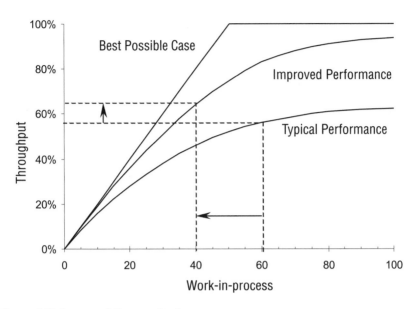

**Figure 5.2b** *Improved Factory Performance*

Fortunately it is possible to change the shape of the factory performance curve (Figure 5.2b). It turns out that the shape of the performance curve is determined by the amount of *variability* in the system. We can see that *factory performance is largely determined by variability.* Higher variability corresponds to poorer performance, and lower variability corresponds to better performance. This highlights the critical importance of reducing variability in today's factory.

What do we mean by variability? The variability to which we refer is the variability of processing time. It can be traced to many internal and external sources such as:

- human operators who follow irregular or inconsistent work procedures
- unreliable equipment that does not always run at the same speed
- differences in material or environmental conditions
- unexpected outages due to power or equipment failure or lack of material
- planned downtime for repair or setup operations, even though the outage is anticipated
- scrap and rework (in severe cases, these can require rescheduling an entire customer order)
- unreliable suppliers, and
- unreliable transportation perhaps due to poor weather and road conditions.

External sources of variability influence processing time by affecting the availability of material needed for production. Internal sources of variability influence processing time directly and affect the availability of material at downstream operations.

Perhaps the most severe form of variability is the variability that we as managers can inflict on the factory. For example, policies of large lot production or batch material movement contribute excessive variability and can easily cause a factory to perform at the worst case level! Higher variability leads to more congestion, more WIP, longer cycle times, higher costs, longer lead times, worse customer service, lower revenue, and lower profits. *Whatever its cause, variability encumbers a factory!*

## Variability Basics

Both the batch-and-queue and single-piece flow examples in the previous section represent *deterministic* systems. By this we mean there is no randomness present. Both systems are completely predictable. However, the batch production example does have considerable variability! Remember, the variability to which we are referring is variability in processing time. The first item processed must wait until all others in its batch are processed before it can pass with its batch to the next station. So, even though it only takes 1 minute to process the first item, from the perspective of the first workpiece it requires 10 minutes. When 10 minutes have elapsed, not one but all items pass to the next workstation. Mathematically this is equivalent to 1 item requiring 10 minutes of processing time, and the other 9 items requiring 0 minutes. This is extreme variability! However, it is neither random nor unanticipated variability; it is variability that results from an inappropriate production policy.

Since processing time variability is such an important parameter for analyzing production systems, we need to have an *objective measure of variability*. For this purpose we introduce the squared coefficient of variation, *scv*. The coefficient of variation, *cv*, is simply the standard deviation divided by the mean. The squared coefficient of variation, *scv*, is the *cv* multiplied by itself. Note that when analyzing production system performance, *time is the critical variable*. (We designate the mean time as *t*. The standard deviation has its usual designation, $\sigma$.)

$$cv = \frac{\sigma}{t} \qquad scv = \left(\frac{\sigma}{t}\right)^2$$

The *scv* of processing time conveys more information about production system performance than either the mean or the standard deviation alone. For example, consider two manufacturing processes. An assembly operation has an average processing time of 20 minutes and a standard deviation of 1 minute. A machining operation has an average processing time of only 2 minutes but the same standard deviation of 1 minute. Clearly, the standard deviations are the same, but the variation in processing time for the machining operation is a much larger proportion of total processing time.

Assembly: $\quad scv = \left(\dfrac{\sigma}{t}\right)^2 = \left(\dfrac{1}{20}\right)^2 = 0.0025$

Machining: $\quad scv = \left(\dfrac{\sigma}{t}\right)^2 = \left(\dfrac{1}{2}\right)^2 = 0.25$

The *scv* for the machining operation is *100* times higher than the *scv* for the assembly operation! Even though the machining operation is ten times faster, it will actually contribute much more variability to the system than the slower assembly operation will contribute.

Almost any manufacturing decision such as lot size, container size, staffing of shifts, machine purchases, maintenance policies, and line configuration affects processing time variability. However, the impact of a particular decision on overall system performance is often not obvious. Fortunately for us, the *scv* provides an objective measure.

Let's practice using the *scv* to make a line configuration decision for a factory. Consider a production operation with an average processing time of 4 minutes for each item produced. The processing times are not deterministic, so they vary randomly; we will assume an *scv* of 0.5. After returning from a conference on lean manufacturing, one of your floor supervisors suggests that the production line could be reconfigured as a cell staffed with five people. Each operator would perform part of the overall production task. The processing time at each station would be 0.8 minutes (48 seconds). Again, for illustration purposes we will assume the *scv* for each of the subtasks is also 0.5. Is the cellular arrangement a good idea from a factory physics perspective?

We can use the definition of *scv* to determine the variance of each individual subtask. Then, by adding the variance of each subtask together (in this case multiplying by 5), we can determine the total variance for the system. Finally, by dividing the variance for the system by the square of the total processing time, we can obtain the *scv* for the total system. This can be compared to the original *scv* to determine which arrangement introduces less variability overall.

$$\sigma_i^2 = t_i^2 \times scv = (0.8)^2 \times (0.5) = 0.32$$

$$\sigma^2_{(5\text{-}stations)} = 5 \times \sigma_i^2 = 5 \times (0.32) = 1.6$$

$$scv_{(5\text{-}stations)} = \left(\frac{\sigma^2_{(5\text{-}stations)}}{t^2_{(5\text{-}stations)}}\right) = \left(\frac{1.6}{4^2}\right) = 0.1$$

Clearly by breaking the large production operation into five smaller subtasks, the *scv* for the line drops quite significantly from 0.5 to 0.1. Factory physics supports the reduction of variability as a worthwhile goal, so the reconfiguration of the production line into a work cell is certainly an excellent idea!

Factory physics can also be used to develop a productive maintenance policy. First we must define the equipment parameter known as *availability*. Essentially availability is the fraction of "uptime" or the percentage of time during which the equipment *is available* to do useful work. Note that availability is not the same as capacity utilization. (Utilization will be discussed later in this chapter and also in Chapter 12.) It is quite reasonable to strive for 100% equipment availability, but it is never appropriate to strive for 100% capacity utilization!

Mathematically availability can be expressed as follows:

$$A = \left(\frac{m_f}{m_f + m_r}\right)$$

where:  $A$ = availability
 $m_f$ = mean time to failure
 $m_r$ = mean time to repair.

Now consider a workstation that operates an average of 70 hours before it must be shut down for maintenance. During the downtime several routine procedures such as changing filters, changing oil, and sharpening tooling are performed. The average time to get the station up and running is about 10 hours. Maintenance personnel support this policy because of its apparent efficiency. They are not bothered very frequently, and when they are needed, they can take the whole day to perform several maintenance tasks. Availability of the machine can be computed as follows:

$$A = \left(\frac{m_f}{m_f + m_r}\right) = \left(\frac{70}{70 + 10}\right) = 0.875$$

This means the machine is available for use 87.5% of the time.

A floor supervisor suggests that the maintenance policy be changed. She suggests that the machine be turned off intentionally every 3.5 hours for 30 minutes of maintenance. Now that you are armed with a growing understanding of factory physics, and you know that factory physics recommends lower variability wherever possible, your first inclination is to reject her suggestion immediately. After all, short, frequent interruptions are a form of variability, right? Nevertheless, you decide to indulge her long enough to run a quick availability calculation.

$$A = \left(\frac{m_f}{m_f + m_r}\right) = \left(\frac{3.5}{3.5 + 0.5}\right) = 0.875$$

Interestingly, her policy provides the same workstation availability as the policy now supported by the maintenance department. At this point it is tempting to regard the two policies as equivalent. However, you still have not proven which policy results in higher variability: short, frequent interruptions or long, infrequent interruptions. To calculate this you need some more information about the workstation, namely the original $scv$ of the workstation and the effective processing time.

Assume the workstation has an average processing time of 12 minutes with a standard deviation of 2 minutes. The $original$ $scv$ for this workstation can be obtained easily as follows:

$$scv = \left(\frac{\sigma_o}{t_o}\right)^2 = \left(\frac{2}{12}\right)^2 = 0.0.28$$

where: $t_o$ = original processing time
$\sigma_o$ = standard deviation of original processing time.

Another important production parameter is *effective processing time*. Effective processing time takes into account the effects of availability and is defined as:

$$t_e = \left(\frac{t_o}{A}\right)$$

In this example, the average processing time is 12 minutes. Availability has already been computed as 0.875, so the effective processing time can be computed readily:

$$t_e = \left(\frac{t_o}{A}\right) = \left(\frac{12}{0.875}\right) = 13.7$$

Although the processing time is only 12 minutes, the long run average, or effective processing time, is 13.7 minutes.

Returning to the question of which maintenance policy introduces less variability into the system, you need to determine the *effective scv* for the workstation under both policies. The effective *scv* takes into account the fact that the workstation does not run constantly and must be repaired periodically. The effective *scv* can be represented by a concise mathematical model if we make some simple and quite reasonable assumptions about the distribution of time until failure and repair time (Hopp and Spearman 1996: 262).

$$scv_e = scv_o + \frac{2m_rA(1-A)}{t_o}$$

Now let's compare the effective *scv* for each of the maintenance policies. First we compute the effective *scv* under the policy of less frequent and longer maintenance procedures (running for 70 hours and maintaining for 10 hours).

$$scv_e = scv_o + \frac{2m_rA(1-A)}{t_o} = 0.028 + \frac{(2)(10)(0.875)(1-0.875)}{12} = 0.21$$

The effective *scv* for this policy is 0.21. (This is actually quite low. Many production managers would love to have their equipment running with such low variability!) Now, to complete the comparison you compute the effective *scv* under the policy of more frequent and shorter maintenance procedures (running for 3.5 hours and maintaining for 0.5 hours).

$$scv_e = scv_o + \frac{2m_rA(1-A)}{t_o} = 0.028 + \frac{(2)(0.5)(0.875)(1-0.875)}{12} = 0.037$$

Clearly the variability associated with shutting down the workstation frequently for short maintenance activities is only a small fraction of the variability associated with the first policy. So, even though it might seem counterintuitive, the supervisor's suggestion of more frequent, shorter maintenance activities is well supported by rigorous factory physics analysis.

Furthermore, since the second term in the effective *scv* equation contains $m_r$ in the numerator, we can make a general statement about such situations: in general, *for the same equipment availability, shorter repair times lead to lower variability*. This is quite reasonable, but it is far from obvious. Most maintenance policies specify running equipment as long as possible before shutting down for repair.

In this example we assume a known average repair time; however, we do not assume that each repair time is a constant. Therefore, our conclusion is valid for both planned and unplanned outages. It is well known that a policy of regular productive maintenance will reduce the occurrence of longer, unexpected downtime.

Factory physics can be applied to other types of manufacturing decisions. Consider a machine that is scheduled to produce in lots or batches because the machine requires a lengthy changeover operation. Lean manufacturing philosophy emphasizes small-lot production. We address the issues of small-lot production in detail in Chapter 11, but two comments are appropriate here. 1) If small lot production is pursued, setup time *may* need to be reduced to maintain sufficient capacity. 2) Reductions in lot size lead immediately to lower variability.

As another example of the application of factory physics, consider a process that generates a large amount of scrap or items that need rework or repair. As expected the *scv* of effective processing time increases as the fraction of rework increases; therefore, more rework leads to higher variability. So, the factory physics focus on reducing variability highlights the need for zero scrap, zero rework, and, in fact, zero defects (Chapter 9).

## Factory Physics and the Value Stream

Optimization of individual workstations is a dangerous goal for today's factory. "Islands of excellence" promise tantalizing benefits, but there is no guarantee that holistic improvement of the factory will result. Today's factory must look beyond the individual or departmental levels in order to optimize the entire value stream. Fortunately, we can use factory physics to help us do this.

As we have mentioned, factory physics is a relatively new specialty within the broader field of operations management, and it relies heavily on queuing theory. (Very briefly, people who go to the bank in England do not wait in lines like we do in the U.S. They wait in queues. Queuing theory is simply the science of waiting.) While queuing theory is not new, its application in industry is still somewhat uncommon. At the turn of the twentieth century, the analysis of *single-station queues* began with Erlang's studies of telephone communications. One application of Erlang's work was the Saturn moon rocket project during the 1960s. Early telephone company computer programs were used to establish (or at least verify) adequate facility size for construction of the rockets and the required number of test bays, barges, docks, and test stands needed to meet the projected flight schedule.

Now, nearly a century after Erlang's original research, we still find *extremely* limited practical application of queuing theory. In fact, a longitudinal study of operations management techniques from 1973 to 1983 shows that queuing theory is used by less than 10% of operations research practitioners. (Most OR practitioners have engineering degrees and are administrators, analysts, engineers, technical specialists, and scientists.) Of the twelve quantitative techniques practitioners list as the most important (such as statistics, probability, linear programming, and simulation), queuing theory ranks tenth (Harpell, Lane, and Mansour 1989: 71)!

Some researchers are beginning to lament the lack of practical orientation in the literature and the absence of queuing theory applications in industry. A deterrent to application is that the articles tend to be highly theoretical rather than practical in the problems they address. This wrongly suggests to some that queuing theory is only a mathematical oddity. "The mathematical and often abstract nature of many articles on queuing suggests that queuing theory is just that: only theory and difficult to apply in practice" (Buchanan and Scott 1992: 29).

The benefits of applying queuing theory are evident even if the instances are limited. Consider the following two axioms from queuing theory. 1) The queue length of a single-server queue with Poisson arrivals is directly proportional to the variance of the service time distribution. This means *it is the variance of service times, not the mean service time, that determines the length of a queue.* 2) A single multiple-server queue with *m* servers is more efficient than *m* single-server queues. This means that one queue leading to many servers is more efficient than multiple queues, each leading to its own server.

The first axiom is partly responsible for the proliferation of express lanes in supermarkets. Our local grocery has a special lane for 10 items or fewer and another lane for 15 items or fewer. All the other lanes are for an unlimited number of items. This allows people who are buying only a pack of gum to wait behind others who are also buying just a few items. The assumption is that a shopper with just a few items can be processed faster than a shopper with two cartloads of groceries. If the shoppers who can be processed quickly go to special express lanes, the variance of processing time for any given lane will be reduced, and all of us will get home sooner.

The second axiom has led to the widespread adoption of single multiple-server queues. A single multiple-server queue is a single queue that leads to a place where many servers are available. This is becoming common in U.S. banks. All patrons wait in the same queue (often designated by velvet ropes). There are several tellers helping customers; the next available teller serves the next person in the queue. The airlines, U.S. Post Office, and even some rides at Disney World® are now using this single multiple-server queue approach! Application of these queuing theory axioms has made life more pleasant. Not only do they reduce waiting time, they are perceived as being fairer!

Now, let's apply queuing theory to production routings and see what insight factory physics gives us about the value stream. First we must define the important term *utilization*. Utilization is the fraction of time the workstation is busy:

$$u = \frac{t_e}{t_a}$$

where: $t_a$ = the mean time between arrivals of production workpieces
$t_e$ = the mean service time or the *effective* processing time.

For stability, the effective processing time must be strictly less than the time between arrivals. Otherwise, both WIP and cycle time become infinite. Put another way, the processing rate must always exceed the arrival rate. This is not a special case; it is a requirement of all production systems!

In order to demonstrate how the variability in one station affects the variability throughout the value stream, we will model the flow of a product from station to station along its routing. Consider a discrete-parts manufacturing system consisting of numerous workstations. The workstations may be automated machines, manual machines, manual assembly stations, or virtually any other type of production process. As the product moves from station to station, it must wait in queue to be processed. Such a manufacturing system can be thought of as a network of queues (Suri, Diehl, deTreville, and Tomsicek 1995: 129).

The waiting time (or cycle time) in a queue can be calculated as follows (Medhi 1991):

$$CT_q = \left(\frac{scv_a + scv_e}{2}\right)\left(\frac{u}{1-u}\right)t_e$$

where: $scv_e$ = the squared coefficient of variation for effective processing time
$scv_a$ = the squared coefficient of variation for time between arrivals.

So, the cycle time in a queue is proportional to the variability of effective processing time *and* the variability of time between arrivals. Also, as a consequence of Little's Law, the WIP in the queue is affected the same way. Furthermore, we can see that utilization dramatically affects the queue time (and WIP). If utilization approaches 100%, both the cycle time and WIP become infinite! (Chapter 12 addresses the pursuit of high capacity utilization and its inappropriateness for today's factory.) As a measure of operational efficiency, utilization is informative to a certain degree. However, it is not a production parameter that should be maximized. Attempting to do so leads to longer cycle time, longer customer lead times, more congestion, excessive WIP, poor flexibility, poor customer service, lower revenue, and lower profit (Buchanan and Scott 1992: 30)!

As might be expected, variability at one workstation affects the behavior of other stations throughout the value stream. An important implication is that reducing variability early in a production routing is more effective than equivalent reduction later in the routing. This is easy to see. The variability in time-between-departures from one station becomes the variability of time-between-arrivals at the next station. The variability propagates and amplifies from one station to the next. Mathematically this is shown in the following equation for the $scv$ of departure time:

$$scv_d = u^2(scv_e) + (1 - u^2)scv_a$$

While this may seem completely abstract, it has important practical implications for material flow throughout the value stream. The variability of time-between-departures from the first station affects the cycle time in queue at the second station just as much as the variability in processing time of the second station does. The equations for cycle time in queue and departure $scv$ can be entered easily into a spreadsheet computer program and used to analyze entire production routings.

Let's look at another simple example. Consider a materials management policy that sends 8 hours worth of material, say 40 pieces, to a workstation at the beginning of every shift. It might seem like an excellent policy from a variability standpoint since all the "jobs" arrive at once. In fact, it is tempting to characterize the $scv$ of arrival times as zero! However, we need to take a closer look. Since 40 pieces arrive every 8 hours, the average time between arrivals is simply 8 hours divided by 40 pieces, or 0.2 hours per piece. The variance of time between arrivals can be computed easily as follows:

$$\sigma_a^2 = \frac{1}{(\# jobs)}(time)^2 - (t_a)^2 = \frac{1}{40}(8)^2 - (0.2)^2 = 1.56$$

Therefore the $scv$ of time between arrivals is

$$scv_a = \frac{1.56}{(0.2)^2} = 39$$

An $scv$ of 1.0 is considered the practical worst case. An $scv$ of 39 represents tremendous variability! An $scv$ of 39 is 39 times worse than the variability expected in a process operating at the practical worst case. Hopp and Spearman (1996: 270) point out a general relationship for batch transfers. If material arrives in batches of $k$ pieces, the $scv$ of time between arrivals is given by:

$$scv_a = k - 1$$

Notice that we do not mention randomness in this analysis of variability. The variability is not an artifact of late deliveries, poor quality, or incorrect shipment quantity. It is not a result of uncertainty in the arrival time. *The extremely high variability is simply a result of moving material in batches.* We can see clearly that batch arrivals contribute significantly to the variability in an entire production routing!

Many factories encounter severe variability because they have a policy of large material transfer batches. Production operators, material handlers, and even managers try to improve "efficiency" through batch delivery of materials. Often batch material movement *is* more "efficient" (meaning that less effort is required) for an individual or a department, but the impact on overall factory performance is devastating. A better approach is to strive for frequent, regular movement of small lots throughout the value stream. Interestingly, this result is completely consistent with lean manufacturing philosophy.

## The Laws of Factory Physics

As we mentioned above, factory physics is an incipient science of manufacturing. It identifies descriptive models from which to build intuition and then establishes useful prescriptive models. This approach has led to several laws and generalities that are known collectively as the *laws of factory physics*. Some of these are inherent truths; in fact, some are even tautological (Little: 1992). Other factory physics laws are true "most of the time" (Hopp and Spearman 1996: 282).

"Most of the time" is not a condition unknown to science. Even the familiar laws of Newtonian physics are only approximations of reality. They hold true "most of the time" when speeds are slow (with respect to the speed of light) and when distances are large (with respect to the diameter of an electron). (Gamow (1968) offers some amusing and quite accurate short stories about how life would be if the speed of light were slow (on the order of 10 mph) and Planck's constant were large (on the order of 1.0).) Even though the Newtonian laws are not entirely accurate, they still hold up our bridges and our buildings without too much trouble. Likewise, the "laws" of factory physics are very useful for running today's factory. We have chosen several of the more profound ones to include here. Readers who would like to know more about these laws are directed to the original work by Hopp and Spearman (1996).

- Little's Law: Throughput = WIP divided by cycle time.
- Capacity: In steady state all factories will release work at an average rate that is *strictly less* than average capacity.
- Variability: In steady state increasing variability *always* increases average cycle time and WIP levels.
- Variability Placement: Variability early in a routing has a larger impact on WIP and cycle times than equivalent variability later in the routing.
- Utilization: If a system increases utilization without making any other changes, average cycle times will increase in a highly nonlinear fashion.
- Move Batches: Cycle times over a segment of a routing are roughly proportional to the move batch sizes used over that segment.

- Process Batches: As process batch size increases, cycle time increases proportionately.
- Pay Me Now or Pay Me Later: If you cannot pay for variability reduction, you will pay in one or more of the following ways: 1) long cycle times and high WIP levels, 2) wasted capacity, or 3) decreased throughput.
- Lead Time: The manufacturing lead time for a routing that yields a given service level is an increasing function of both the mean and variance of the cycle time for the routing.
- Cycle Time: The average cycle time for a routing is the sum of the average cycle times for the stations on the routing. For each station the average cycle time = queue time + process time + wait for batch time + move time (+ wait time for other components in an assembly operation).
- Self-Interest: People, not organizations, are self-optimizing. Optimizing individuals' jobs does not optimize the factory.
- Responsibility: Responsibility without commensurate authority is demoralizing and counterproductive.

## Summary

There has been a conspicuous absence of operations research applications in industry. Practical applications have been discouraged by the fact that most research focuses on highly specific results rather than general principles. Fortunately this is changing. One example is the application of queuing theory. "We are seeing that the strategic impact of queuing models in manufacturing can be Herculean" (Suri, Diehl, deTreville, and Tomsicek 1995: 129). Furthermore, researchers are beginning to discover basic "laws" of manufacturing (Little 1961; Askin and Standridge 1993; Buzzacott and Shanthikumar 1993; and Hopp and Spearman 1996). These laws can be used to make sound manufacturing decisions that are based on scientific principles.

Factory physics provides a way to predict the ramifications of manufacturing policies, and, consequently, better policies can be made. Managers can be more confident about making decisions when they are founded in sound scientific reasoning. The resulting policies are much easier to implement than policies based on imitation or anecdotal case studies.

The examples included in this chapter are intended to build manufacturing intuition and confidence. It is not always obvious how a specific action or decision in one part of the factory will affect the overall production system, and sometimes conventional wisdom can lead us astray. Factory physics can eliminate some of the guesswork by providing analytical tools to help us make tough manufacturing decisions.

Batch and queue production leads to worst case factory performance. Single-piece flow yields best case performance. For years we have comforted factory managers by stating that all factories operate somewhere in between these two extremes. This is not always the case. A "work cell" can exhibit worst case performance if material is moved in batches through the stations in the cell. A factory can perform worse than the theoretical worst case if material moves by *batch and store* or *batch and wait* rather than true batch and queue.

Factory physics warns that reducing inventory without making other improvements leads to a decrease in throughput, revenue, and profitability! Nevertheless, reducing inventory is the most common reason factories apply the just-in-time (JIT) techniques. It is true that a lean factory operates successfully with much less inventory, but low inventory is not the reason the factory runs well. Low inventory is simply a result of using lean strategies to run today's factory effectively.

*A lean factory runs well because variability has been reduced.* Both lean manufacturing and factory physics call attention to variability. Schonberger (1986: 14) refers to variability as "a universal enemy!" Variability can come in many forms: natural variability, quality-related variability, machine outages, management decisions, and external events. Low variability allows the factory to have shorter cycle time, shorter customer lead time, greater flexibility, less inventory, better utilization of workers and equipment, higher revenues, and ultimately higher profit. It is variability, not inventory, that today's factory must reduce. The very challenging task of running today's factory is made much easier when variability is reduced wherever possible.

It is interesting to note that in case studies the *factory physics analysis leads to the same conclusion we would have reached using lean thinking*. We have found this to be true in general. When factory physics is applied to a manufacturing situation, we arrive at the same decision we would have made using lean philosophy. This has prompted some researchers to call factory physics the *science of lean manufacturing* (Duenyas 1989). It is important to remember that the two philosophies originated independently and quite differently. Lean philosophy developed out of necessity in post-WWII Japan and was seeded by developments from early American manufacturing. Factory physics grew out of operations research and queuing theory and is firmly rooted in rigorous mathematical analysis. Their dichotomous origins make their mutual corroboration all the more impressive. Together they provide powerful analytical and philosophical resources for running today's factory.

*Chapter 6*

# Manufacturing as a Management Strategy

## Introduction

In a business context, *goal* and *strategy* are sometimes used interchangeably. In our discussion we make a distinction between these terms for reasons of clarity and consistency with operations research literature (Ward, Bickford, and Leong 1996). A business or *corporate goal* is the explicit, desired outcome a company is trying to achieve. Perfect quality, a 30% market share, and overnight delivery of goods or services are examples of corporate goals. A business or *corporate strategy*, on the other hand, is the approach a company takes to achieve its business goals. Corporate strategy, sometimes referred to as *competitive strategy*, is an explicit plan that outlines steps for accomplishing long-term goals. The corporate goals are where the organization is going, and the corporate strategy is how the organization will get there (Daft 1992: 42).

Just as a company has goals and strategies, today's factory needs goals and strategies as well. In other words, the factory needs a *manufacturing strategy*. What do we mean by manufacturing strategy? *Manufacturing strategy is a pervasive theme or mission that guides decision making, establishes direction, and determines what competencies the factory will develop.* It is a long-term roadmap for operations management (Schroeder, Anderson, and Cleveland 1986). Manufacturing strategy can be considered a "blueprint" for how the factory should be organized, what equipment should be acquired, what production goals and measures should be adopted, and what capabilities the factory should develop or eliminate (Bates, Amundson, Schroeder, and Morris 1995; Hayes and Pisano 1994).

A coherent manufacturing strategy promotes a consistent pattern of decision making (Hayes and Wheelright 1984): factory management can make decisions, investments, and policies that support the business goals of the company. A factory's manufacturing capabilities are not easily duplicated; when those very capabilities help define the business goals and strategies of the company, the factory itself becomes a powerful and sustainable source of competitive advantage.

So, in summary, there is a complex and dynamic interrelationship among the corporate goals, corporate strategy, manufacturing strategy, and manufacturing capabilities. Corporate goals are the focus of the corporate strategy. Corporate strategy influences the manufacturing strategy. Manufacturing strategy defines the long-term capabilities of the factory. The factory's capabilities, in turn, help shape the corporate goals.

The idea that *what the factory is doing* should be aligned with *what the company is doing to reach its goals* is certainly not new. Since the time of the early artisan workshops (Chapter 2), manufacturers have established business strategies that were compatible and consistent with their manufacturing capability. Nevertheless, to our knowledge the subject of manufacturing strategy was not discussed explicitly in the business, academic, or manufacturing literature until 1969 when Skinner published his landmark article, "Manufacturing—Weak Link in Corporate Strategy." So, for at least thirty years researchers and manufacturing managers have been *explicitly aware* of the importance of aligning manufacturing strategy with the overall corporate strategy.

In light of this discussion, it is surprising that, at the close of the 20th century, only about 10% of manufacturing companies actually have a manufacturing strategy (Ettlie and Penner-Hahn 1990). One plausible explanation for this lack of strategy is that many top managers are simply more interested in marketing or finance than in manufacturing (Nicholas 1998: 17). Some corporate managers consider the factory nothing more than a cost center—a necessary evil that must be endured in order to obtain products to sell. "Most companies seek to minimize the possible negative impacts of their manufacturing operations . . . They ignore the fact that manufacturing can be a competitive weapon, rather than just a collection of ponderous resources and constraints" (Gupta and Somers 1993: 87). So, we find that most companies do not have a manufacturing strategy; those that do, often have strategies that are misaligned with the overall corporate goals.

If manufacturing policies are inconsistent with corporate strategy, the obvious repercussion is a deterioration of corporate performance (Gupta and Somers 1993: 88). Without common direction, manufacturing policies or decisions in one area of the company may not be synergetic with those of other areas. Resources may be committed to projects that do not help the company reach its ultimate goals. Furthermore, if manufacturing is decoupled from corporate strategy, manufacturing policies and decisions may contradict and even subvert the overall corporate goals (Gupta and Lonial 1998: 246).

This chapter discusses manufacturing strategy and offers guidelines that enable managers to establish consistent manufacturing strategies that are well aligned with corporate objectives. We also present a critical thinking approach to use when choosing manufacturing initiatives. Lead time reduction is discussed as a manufacturing strategy that supports several excellent business goals, and we offer several practical ways to reduce lead time in a manufacturing operation.

## The Six Facets of Manufacturing Strategy

While manufacturing strategy is critical to the success of today's factory, it is an admittedly abstract construct. A useful way to observe and understand manufacturing strategy has been proposed by Leong and Ward (1995). They identify six facets of manufacturing strategy: three of these pertain to the content of the manufacturing strategy, and three pertain to the process through which the strategy is put into effect. We discuss these six facets of manufacturing strategy so our readers can more easily discern and develop their own manufacturing strategy. The six facets are planning, proactiveness, patterns of actions, portfolio of manufacturing capabilities, programs of improvement, and performance measures.

Most commonly, manufacturing strategy is seen as a factory's long-term *plan*, and, in fact, that is the aspect we have presented so far. This planning aspect of manufacturing strategy does not necessarily need to be driven from the top down; however, in companies that have not yet reached *world class* status, the "plan" is primarily the concern of senior management. The planning aspect is especially useful during major changes or large capital spending programs. While this planning is certainly a valid aspect of manufacturing strategy, it is also incomplete.

Developing capabilities before they are needed is the *proactive* facet of manufacturing strategy. We indicated earlier that manufacturing strategy directs what capabilities the factory will develop; anticipating the potential of new technologies and new capabilities allows manufacturing to be an important part of product development and marketing. Leong and Ward present an example of this proactive facet of manufacturing strategy from the electronics industry. In the early 1980s, the potential for surface mounting the components onto circuit boards promised smaller, lighter, and more attractive products. It also required gluing instead of soldering as an attachment method thereby rendering rework much more difficult and costly. Companies that developed this capability were in a position to design and market products that exploited those capabilities. They could introduce new products faster and with higher profit margins than reactive manufacturers could.

Another facet of manufacturing strategy is the *patterns of action* that can be observed over time. We make an important distinction between patterns of decisions and patterns of actions. Patterns of decisions indicate the intentions of management or the intended strategy, not necessarily the strategy that really exists. Patterns of action are observable, even when they are unintentional or when they run counter to the "official" strategy. Consider a company with a stated manufacturing strategy that includes a transition to lean manufacturing. Yet, the actions of middle managers and shop floor workers continue to support overproduction, producing in large lots, keeping large quantities of WIP in-between processes, and moving production material from station to station in large batches. The pattern of action that emerges is not consistent with a lean manufacturing strategy. We conclude, therefore, that this company has a mass production manufacturing strategy regardless of official statements to the contrary.

Since manufacturing strategy directs what capabilities the factory will develop, it is not surprising that one facet of manufacturing strategy is a *portfolio of manufacturing capabilities*. This portfolio includes five generic manufacturing capabilities: cost, quality, delivery performance, flexibility, and innovation. Excellent manufacturing companies continue to add and increase their capabilities; as excellence in one area is achieved and maintained, emphasis shifts to another capability. This makes it extremely difficult for competitors to catch up.

Another widely publicized facet of manufacturing strategy is a *program of improvement*. This is a particularly revealing aspect of manufacturing strategy. By examining what the factory seeks to improve, we gain considerable insight into the nature of the manufacturing strategy. Obviously, the improvement programs should be improving things that are strategically important; however, even if the improvement programs are focused on "nonstrategic" aspects of production, they reveal a strategy in action that is misaligned with top management's "official" plan. For example, a manufacturing strategy may emphasize reducing lead time to support the official business goal of quick response to customer demand (speedy delivery). If the factory institutes a cost reduction program that results in longer customer lead time, the "improvement" effort reveals that the true manufacturing strategy is not customer service oriented.

Performance measures are the sixth facet of manufacturing strategy. The measurement criteria should reflect the strategic goals of the factory, and the reward system should be aligned with the manufacturing strategy. If they are aligned, then it is obvious what actions and behaviors are of strategic importance. If they are not aligned, which is often the case, people's actions and behaviors will be influenced by the reward and measurement system regardless of the "official" statement of strategy. The performance measurement system reveals candidly the true manufacturing strategy.

Manufacturing strategy is not a difficult concept to understand, but it is difficult to actually observe. The six facets presented here can be interpreted as windows for viewing manufacturing strategy. The view through any one window is necessarily skewed and incomplete. Several views must be used and integrated to give an accurate picture of a company's manufacturing strategy.

## Lean Manufacturing as a Management Strategy

In the autumn of 1997, the first annual *Blue Ribbon Panel on Global Manufacturing* met in Traverse City, Michigan. The purpose was to define manufacturing strategies that could guide global manufacturers into the new millennium. The panel was comprised of senior manufacturing executives from around the world. They suggested that *"companies should focus on lean methods of manufacturing and continuous improvement practices"* (Aldred 1998b: 6). Other key conclusions by the panel were that single-piece flow is the optimal manufacturing method for meeting future customer demand, that customers should pull product through the value stream, and that lean manufacturing is the best hope for a common global language of manufacturing.

At present, the largest and most influential manufacturing companies in the world are incorporating lean thinking into their operations management. They are attempting to transform their operations in order to realize the tremendous benefits associated with lean manufacturing strategies. Many of these leading companies are creating their own production systems patterned after the now famous

Toyota Production System. Too often, however, they are copying the practices without understanding the fundamental principles.

We are not the first to notice the striking discrepancy between *imitation* and *understanding*, nor the first to caution against copying practices without understanding fully the principles. Robert Inman warns: "Without understanding the production system, it's easy to misapply JIT techniques" (1993: 44). He goes on to restate the admonitions of Taiichi Ohno who cautioned that "a little knowledge is dangerous." Shingo gives a similar warning: "In addition to an appreciation of the techniques of the Toyota production system, an understanding of the concepts that lie behind those techniques is crucial. If it is missing, errors in application are unavoidable" (1989: 213). Deming makes a more generalized point: "It is a hazard to copy. It is necessary to understand the theory of what one wishes to do or to make" (1989: 5).

From the information we have presented so far, it should be clear that imitation without understanding does not yield a sustainable competitive position. In fact, there is growing evidence that suggests imitation may not provide any advantage at all. "Imitation strategies may be even less effective than standard policies" (Ocana and Zemel 1996: 212). "In some cases, the best-practice strategy can worsen a firm's situation as it selects practices unsuited to its operating environment" (Pilkington 1998: 40).

While it is dangerous to imitate a "lean" company and apply "best practices" indiscriminately, it is even more dangerous to modify "lean principles" to fit an existing corporate culture. Doing so simply results in a new name for the same anachronistic production strategies that led to noncompetitive performance in the first place. Running today's factory should not involve mimicking other manufacturing companies and applying their solutions to unrelated problems. *Running today's factory requires understanding the strategic business goals of the company and developing a manufacturing strategy that supports those goals.*

## Developing a Manufacturing Strategy

We emphasize that manufacturing should be strategic. Changing a factory *just because we can* does not justify the difficult transformation journey. We must have some compelling business reason for the factory to run differently. For example, the factory might be losing market share, or its product quality might not be adequate. Perhaps the lead times are too long, or the prices are too high. Perhaps the owners are not satisfied with the returns they are getting, and they want the factory to generate more profit. Perhaps there is increasing competition from offshore companies with low production costs and low prices. Any of these concerns could translate into corporate goals, which, in turn, would influence the manufacturing strategy.

When establishing corporate goals, we recommend keeping a strong focus on the customer. *What is most important to the customer?* Satisfying the customer is the only way the factory can succeed. "After all, satisfying the customer is what business is all about (*The New York Times*, 23 February 1998). If the customer is not satisfied, it does not matter how efficient the operation is, how wonderful the quality is, or how short the lead time is. Since the factory must support the corporate objectives, establishing a manufacturing strategy requires first understanding the corporate goals, preferably from the perspective of satisfying the customer.

## *Assessing Today's Factory*

Before the manufacturing strategy is developed, management should take a close and objective look at the factory. Before we can begin any journey, we must know not only where we are going but also where we are. Similarly, it is difficult to transform a factory without understanding its current state, and this is accomplished through a factory assessment. First we observe objectively *how* the factory is running. Only then can we understand *why* things are running that way. For example, let's say our assessment reveals an inordinate amount of WIP throughout the factory. Why does the factory keep excess inventory? Is equipment too unreliable? Are setup times too long? Is production scheduled in large batches? Is there any reason at all for having the excess inventory? Once we understand the reasons *why*, we can begin to develop appropriate strategies.

There are various ways to assess the factory, and the method used is entirely discretionary. Two observational tools that are often useful are the production flow diagram and value stream mapping. These tools help managers "see" their factories objectively. A production flow diagram is particularly useful in identifying non-value-adding steps in production. The value stream map probes deeper and identifies specific details of production. Consider for a moment drawing a map to your house. During the process of drawing the map, you may find that you do not know all the street names, how many traffic lights there are, or even how many houses are on your block. Yet, each of us (hopefully) is able to find our way home, and we *think* we are familiar with our neighborhood! Furthermore, those of us who have walked home can verify that the level of detail we observe while walking is very different than when we drive.

Similarly, these mapping tools force us to take a more detailed and thorough look at the factory and allow us to see it more objectively. The mapping process helps us observe operations in a way that cannot be done simply by walking across the plant to the cafeteria. By mapping the value stream we notice WIP, nonfunctional equipment, wasted operator motion, poor layout, long setup times, large batches, uneven work loads, and many other production anomalies. These mapping tools allow a complete stranger to walk through a factory and understand the material flow and resource utilization. Regardless of the assessment technique used, the objective is to understand current factory operations with respect to the stated corporate goals. Once the goals have been established and the current situation is clearly understood in relation to those goals, the management is ready to define its manufacturing strategy and develop a transformation plan.

All of us have a difficult time looking objectively at familiar situations. This is part of being human. From infancy we develop basic assumptions about the environment and the way people interact with each another. These assumptions become almost "hard-wired," and it is very difficult to perceive familiar events from a new perspective. Even anthropologists conduct field research in societies with cultures very different from their own. In these "foreign cultures" it is easier to be objective and attentive because many of the scientist's basic assumptions about the world do not apply. Very little is taken for granted, and everything is noted. The smallest details are considered clues, and nothing is accepted without question. The same objectivity and questioning attitude are important when assessing a factory; it is necessary to understand how the factory really runs and not how we assume it runs.

John Buzacott, a leading researcher in the field of operations management, goes so far as to suggest that manufacturing research should be conducted by a team comprising members trained in the empirical social sciences as well as the physical sciences (1995: 125). Fortunately, most factories can be analyzed without conducting a university research project with physicists, anthropologists, and operations researchers. Nevertheless, the point is clear. The team that analyzes the factory should adopt the attitude of an objective scientific researcher, and the investigation should be carried out accordingly. It is important for the assessment to reflect reality rather than assumptions.

Outside consultants are often able to provide an objectivity that full-time employees of the factory are understandably not able to assume. Because they are not immersed in the factory "culture," consultants can be more impartial and unbiased. They may be able to discern subtle or even obvious details that have grown invisible to factory workers and managers. We caution our readers, however, that care should be taken when choosing outside assistance. Bergstrom warns, "Find experts who know more than just mapping value streams and realigning them" (1995: 32).

### Manufacturing Initiatives

Consider the following example. Batchmore Manufacturing Company produces a variety of different items. Its competitive strategy highlights speedy delivery and quick response to changing customer demand. Specific corporate goals are:

- shorter order-to-delivery time (lead time) for common items

- extremely high (99%) customer service (by this we mean the probability of an order being shipped late is less than 1%)
- flexibility to produce many different items and to respond quickly to changing customer demand
- lower investment in inventory
- higher profit.

The Batchmore management team met to discuss ways to improve the manufacturing operations. During the meeting several initiatives were suggested.

(a) The creation of a quality assurance department with a laboratory, staff, and inspection equipment including a coordinate measuring machine and a laser measuring system.
(b) A *kaizen* event to reduce setup times.
(c) A maintenance policy whereby equipment is shut down for 24 hours every two weeks for maintenance.
(d) The implementation of an advanced MRP-II system to replace the outdated MRP scheduling system.
(e) An immediate reduction in WIP.
(f) An immediate reduction in lot size.

Before choosing any of these improvement initiatives, Batchmore's management conducted a detailed factory assessment. The assessment revealed the following:

- chronic equipment problems and frequent downtime
- long setup times
- low machine utilization
- large amounts of work-in-process material (WIP) stored throughout the factory
- the current MRP system schedules production in large lots.

Based on the results of the factory assessment, Batchmore's management met to determine whether or not the suggested improvement initiatives would help them meet their stated corporate goals. You were invited to help. Given the corporate goals and the description of the current situation, which (if any) of the suggested manufacturing initiatives directly support(s) the corporate goals?

Clearly any of these manufacturing initiatives could be worthwhile depending on the factory's current situation. However, the purpose of this exercise is to determine which, if any, of these choices *supports directly the stated corporate goals* given the limited information about the factory's current state.

Choice (a), a quality assurance department, is a tempting one, especially in light of the general trend toward higher quality. However, the current-state factory assessment does not mention any quality deficiencies, and the corporate goals do not indicate that improved quality is necessary. Therefore, you conclude that this initiative does not support *directly* the stated corporate goals.

Choice (b), setup time reduction, is another likely candidate. After all, setup time reduction is one of the few initiatives endorsed by both mass producers and lean producers. In this case the assessment indicates that the equipment utilization is low. Setup time reduction for its own sake is not justified unless capacity utilization is extremely high. Shorter setup time does lead to slightly shorter lead time, but the improvement is usually inconsequential. Therefore, you conclude that this initiative does not support *directly* the stated corporate goals.

Choice (c), regular maintenance, looks like a sure winner. A regular maintenance policy will certainly reduce the unplanned downtime, and the factory assessment states clearly that Batchmore has chronic machine problems. Before you accept this initiative, you look a little deeper. The difficulty with this choice is not with regular maintenance but with the specific policy that was described. Shutting down the equipment infrequently and for such a long time might be efficient for the maintenance department, but it wreaks havoc on the shop floor. The disruption propagates to all subsequent operations necessitating a large amount of WIP. This increases lead time and decreases the factory's ability to respond to changing customer demand. A policy of more frequent maintenance would be more advantageous than the stated maintenance policy. Therefore, you conclude that this initiative is not appropriate.

You reject choice (d), MRP-II, on general principle. The upgrade is not necessarily a bad idea, but it does not support *directly* the stated corporate goals. Even though it may use a more sophisticated scheduling algorithm, it is still a push system. It still generates a schedule, and it attempts to control throughput by regulating the input to the system. Even when operating at its optimal point, such a system is always less profitable than a comparable and optimized pull system (Chapter 9). Therefore, if the factory wishes to improve its production control, a pull system would be a more profitable choice. A pull system provides shorter lead times and greater flexibility to produce whatever the customer wants since there is a marketplace of ready items at all times.

Choice (e), WIP reduction, might be especially tempting to those who read the popular JIT literature. It is true that lean companies with short lead times and the ability to respond quickly to changing customer demand have far less inventory than their mass production competitors. However, *low inventory is a result, not the cause, of their excellent performance*. Reducing inventory without making other improvements will cause longer lead time and worse customer service—the exact opposite of the stated corporate goals. Sadly, many manufacturing companies choose this initiative first. However, you resist this temptation and reject choice (e) as well.

Choice (f), smaller lot sizes, is a rather annoying initiative. It would surely irritate managers whose current departmental performance measures are probably based on standard labor variance. Smaller lots require more frequent changeover operations, and that means less productive output. This initiative could also aggravate the production workers, especially if the setup operations are difficult or unpleasant. Nevertheless, the factory assessment showed excess equipment capacity because machine utilization is low, so this initiative is feasible and should be considered seriously. Smaller lots lead irrefutably to lower variability, lower inventory, lower production cost, shorter cycle times, and shorter lead times. Smaller lots allow more flexibility and quicker response to changing customer demand. These are in precise agreement with the stated corporate goals! Therefore, you accept this as an appropriate manufacturing initiative. The challenge is to find ways to make it less objectionable to the middle managers and workers.

The only suggested initiative that supports unequivocally the stated corporate goals is small-lot production. This example is intentionally simple, but it demonstrates that some very common "improvement" activities do not necessarily help a company reach its business objectives. Also, our conclusion about small-lot production is generally true. Small-lot production often leads to marked and holistic improvement in factory performance.

In the Batchmore Manufacturing example, some corporate goals seem to conflict with one another. For example, if we begin quoting shorter lead times, the customer service will decrease accordingly. Likewise, if we reduce inventory and hold the customer service level at 99%, the quoted lead times will necessarily increase. While the goals may seem mutually exclusive, lean thinking and factory physics help us see how they interrelate. By understanding the relationships, we find that today's factory can actually realize these goals simultaneously. Let's look at the Batchmore case from the lean thinking and factory physics perspectives.

Both lean manufacturing and factory physics emphasize that *pull systems* are inherently more flexible, require less inventory, and are more profitable than scheduled systems. These benefits are certainly aligned with the company goals of flexibility, lower inventory investment, and higher profit. Factory physics also teaches that by *reducing variability* throughout the production stream, inventory can be reduced and customer lead time can be shortened while still maintaining excellent customer service. These benefits are consistent with the stated goals of shorter lead time, high customer service, and low inventory investment.

The Batchmore Manufacturing example demonstrates the importance of thinking critically and examining manufacturing choices in terms of how they contribute to the overall company goals. However, picking an appropriate manufacturing initiative off a list is not the best way to develop manufacturing preeminence. It is far better to develop an integrated and systemic manufacturing *strategy* that supports the corporate goals.

## The Importance of Cycle Time

Interestingly, there is a single manufacturing strategy that provides simultaneously a variety of excellent competitive advantages. This manufacturing strategy is *to shorten cycle time*. Short cycle time provides benefits for the entire corporation, and it is one of the few strategies on which both the sales and manufacturing departments can agree (Hopp, Spearman, and Woodruff 1990: 78). The advantages of short cycle time include the following:

1. Shorter lead time (faster delivery to the customer).
2. Fewer canceled orders and less impact on the factory when orders are canceled.
3. Less reliance on forecasts about future demand.
4. Greater manufacturing flexibility to respond to changing customer demand.
5. Higher quality due to less opportunity for items to be damaged, and a shorter time between making a defect and detecting the defect.
6. Better customer service, which means the probability of an order being shipped late is reduced.
7. Lower inventory, which means less investment, lower carrying costs, lower handling and management costs, less tracking and storing, and less chance of loss due to obsolescence or spoilage.
8. Fewer disruptions due to changes in product design.
9. Easier management of the factory because there are fewer shop orders in the system.
10. Less need to expedite special rush orders.
11. Higher profits due to lower costs and higher revenues.

Since short cycle time is so important to the factory, we should understand *categorically* how to reduce it. Before we discuss strategies for reducing cycle time, let's briefly review three important aspects of cycle time. First, cycle time comprises processing time, setup time, "wait-for-conveyance" time, conveyance time, queue time, "wait-for-material" time, and many other waiting times. By far the largest contributors to long cycle time are queue time and waiting time. Therefore, reducing queue time and waiting time is an excellent way to reduce cycle time. Second, according to Little's Law, cycle time and WIP are proportional to one another for a given rate of throughput. This means we can find processes that are contributing to long cycle time by looking for areas in the factory with a large accumulation of WIP. Third, lead time is a function of the *mean cycle time* and the *variance of cycle time.* The mean is the average, and the variance reflects the difference between longest and shortest cycle times. The average amount of waiting inventory in a production system increases linearly with the standard deviation of cycle time and does not depend on mean cycle time at all. This means that *waiting inventory cannot be reduced simply by speeding up individual processes* (Hopp, Spearman, and Woodruff 1990:78).

# Guidelines for Reducing Lead Time

We have emphasized that generic solutions are dangerous, but a template or *guideline* for finding solutions can greatly simplify the search. This section provides five useful strategies for reducing cycle time and, thereby, lead time in today's factory.

## *Look for Inventory, the Flower of All Evil*

Contrary to the popular JIT literature, inventory in itself is not the "root of all evil." Inventory, and in particular work-in-process inventory (WIP), is merely an outward manifestation of other problems in the factory. *A high WIP level indicates improper conditions, even when those improper conditions are not visibly disrupting the factory.* It is like the flower of a dandelion bearing witness to a plant with a deep and stubborn root system (Inman 1993: 42). Mowing the dandelion down to ground level or popping off its yellow flower may improve temporarily the appearance of a lawn, but it doesn't get rid of the dandelion. Similarly, getting rid of inventory does not get rid of the problems that caused the inventory to collect in the first place.

An excellent first step toward reducing cycle time is to look for inventory, the telltale sign of production problems. Once we find excess inventory, it is dangerous to just remove it. A reasonable assumption is that the inventory is there for a reason. Removing the inventory without addressing that reason would undoubtedly lead to lower throughput and longer lead time. Instead, we can use the excess inventory to find production areas that need attention. Here are ten common reasons why inventory accumulates in a factory.

1. Setup time is too long.
2. Production is scheduled in large batches.
3. The production equipment breaks down frequently.
4. The shop order has been canceled or preempted.

5. Components needed to complete the job are unavailable.
6. Different parts of the plant work different shifts.
7. The next process is working at a slower rate.
8. The next process is working on a different order.
9. The factory is not making what the customer wants.
10. A specialized worker is absent. (No one else can do the job.)

Addressing problems such as these will help the factory reduce its cycle time and lead time. As a consequence of making these improvements, the WIP levels can be reduced without detrimental consequences.

There has been a long debate on the issue of inventory reduction. Should production problems be solved before inventory is reduced, or should inventory be reduced before the problems are solved? The first approach, solving problems first, has the advantage of being risk-free. This minimizes the disruptions that would occur if inventory were reduced first. However, many production problems are difficult to recognize, and trying to find and solve all of them can be time consuming and expensive. Furthermore, this approach postpones the reduction of inventory and the associated benefits. The second approach, lowering the inventory immediately, disrupts the factory. However, previously hidden problems are revealed, providing the factory with an opportunity to solve them permanently. The factory receives more quickly the benefits associated with lower inventory.

We will use the familiar analogy of a lake with a rocky bottom to illustrate these two views. In Figure 6.1 water represents inventory, and the rocks represent production problems. The factory is a boat sailing across the water.

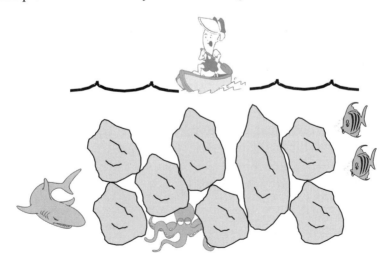

*Figure 6.1* *A Sea of Inventory Can Hide Dangerous Problems*

As we pointed out in Chapter 5, inventory can be seen as a buffer that protects the factory against production problems and *variability*. Most factories attempt to keep enough inventory to prevent common problems from causing too many disruptions. This costly measure has the unfortunate side effect of *hiding* the problems. For example, a factory might keep a large amount of doodads because the machine that makes doodads is unreliable or has a long setup time. The high inventory protects against a stock-out of doodads, and the problems associated with the machine are hidden. Even though the problem is hidden and stock-outs are being prevented, all other negative effects of the problem, such as increased labor effort, longer cycle times and lead times, poor flexibility, and high cost, are still felt by the factory.

The common practice of using inventory to hide problems is analogous to a boat sailing on a deep lake. The water level (inventory) is high enough so that the boat does not hit too many rocks (problems). As water is drained, the boat will begin to hit rocks that protrude from the water. When a rock (problem) is encountered, it should be removed, and the water (inventory) level should be lowered again. Just as lowering the water level makes the rocks apparent to the boat captain and everyone else on board, lowering the inventory level makes problems apparent to everyone in the factory. Many lean proponents advocate this method for finding and solving problems.

The other approach is to use "sonar" to find the rocks, remove them, and *then* drain the water. Sonar provides the same performance improvement as draining the lake but with far less disruption. If this approach is chosen, we suggest a five-step improvement plan.

1. Look for inventory.
2. Understand that it is there to protect the factory from a problem.
3. Find the underlying problem.
4. Solve the problem.
5. Lower the inventory.

This improvement plan should be repeated, continually driving the factory ever closer to perfection.

Which approach is optimal? Recently there has been some excellent analytical and empirical research on this subject (Ocana and Zemel 1996). The research proves that *if the organization is capable of learning from its mistakes*, then lowering the water and hitting the rocks is actually optimal. Hitting the rocks leads to faster improvement because the problems become evident and must be addressed immediately. Unfortunately, most organizations and, in particular, most factories do not learn from their mistakes. They hit the same rocks over and over, time and time again. This may be because removing the rock is difficult, costly, or has been given a low priority. Nevertheless, the result is the same: the rocks (problems) are still there.

We often hear comments like, "Oh, that happens all the time" or "It's been like that for years." These comments are not indicative of a learning organization. Today's factory does not tolerate repeated occurrences of production problems. Today's factory records problems as they occur and establishes systems by which the root causes are identified and solved permanently. The problems are "blown right out of the water!" If a factory is incapable of learning, the "drain the pond" approach is very dangerous. The challenge is to help our factories become *learning organizations*. Factories that are capable of learning from mistakes enjoy a competitive advantage that cannot be easily copied by rival organizations.

### *Keep Production Material Flowing to the Customer*

Cycle time, lead time, and inventory will all decrease if production material is always flowing toward the customer. Recall that the single largest component of cycle time is waiting time. Production material waits in queues, waits for other parts, waits to move, and waits for the remainder of the batch to be finished. If the "non-value-added" waiting time is reduced, then cycle time and lead time will decrease.

One excellent way to reduce waiting time is to make a distinction between production lot size and transfer lot size. Process batches should be split into as many transfer batches as the factory can handle feasibly. This point is missed by many researchers and factory managers. In fact, an award-winning article (Cook 1994) *misstates* that in just-in-time manufacturing the transfer batch size must equal the production batch size. This is totally incorrect! There is absolutely no reason that transfer batch size must equal the production batch size. Forcing the entire lot to wait until the last piece is finished is a terrible waste and contributes significantly to long cycle times.

Another excellent way to reduce waiting time is to establish a "cap" or a maximum on the amount of WIP allowed in the system. This can be accomplished quite easily by instituting a pull production control system (Chapter 9). Limiting the amount of WIP that is present in queues also limits the cycle time (Little's Law). Again, the WIP levels should be set with care to ensure that throughput does not suffer. A pull system is a simple way to reduce cycle time and lead time.

### *Synchronize Production*

It is common for an operator to produce an item he enjoys making regardless of whether the downstream operator needs that particular item. It is also common for a departmental supervisor to overbuild an item that is running well. This avoids a setup operation and increases production output for the shift. Managers often schedule a large production run of "easy" parts early in the month to "bank" the revenue when those parts are received into inventory. The problem jobs are saved until the last week of the month, and the factory appears to run smoothly for three out of four weeks! This makes the plant or department *appear* efficient and profitable, but it is only an illusion. These behaviors can be rationalized from an individual perspective, but they all serve to increase cycle time, lead time, and overall production costs. On the other hand, if production is spread evenly throughout the month, the factory might appear to be having problems all month long. Even though this approach might cause short-term departmental performance measures to deteriorate, the overall operation will benefit.

To achieve holistic improvements, the various production processes throughout the value stream must be synchronized. One way to synchronize production processes is to group them in a manufacturing cell (Chapter 8). Within the cell each worker produces only what is required by the next process. This reduces cycle time in two ways: 1) synchronization is ensured, and 2) single-piece flow production exhibits the shortest possible cycle time (Chapter 5).

Another way to synchronize production processes is to implement pull production control. A pull system ensures that upstream processes produce only what has been consumed by downstream processes. In this way the production processes throughout the value stream are linked and synchronized. Another important secondary benefit is faster detection of defects before they can accumulate.

### Keep the Workload Steady

In some factories, work is released in large batches with shop order quantities in thousands or even millions of pieces. This sets off a series of events that can cause cycle time and lead time to soar. First, the sudden release of work leads temporarily to extremely high resource utilization, which is followed by a lull in activity. It is well known that high workloads cause long cycle time and high WIP (Chapter 12); however, *uneven* workloads cause both the cycle time *and* variability in cycle time to increase (Hopp, Spearman, and Woodruff 1990: 81). If management attempts to sustain the high workload (high resource utilization), the result will be severe congestion, an explosion of WIP, and incredibly long lead time. If management does nothing, a "bubble" of WIP moves through the factory intermittently overloading and starving resources along the way. We have seen this phenomenon many times in actual factories and have come to refer to it as the *pig-in-a-python effect*.

If smaller orders are released more often, the factory resources are loaded much more evenly. Small, frequent releases yield short queues, maximum flexibility, short cycle time, and, consequently, short lead time. This is analogous to the python swallowing dozens of little piglets instead of one large pig. Releasing small orders frequently eliminates the pig-in-a-python effect and provides a smooth, steady workload throughout the value stream.

Surprisingly, many factories prefer to "stretch the python" so it can swallow an even larger hog! (While this analogy might seem a bit outrageous, the famous Amazon explorer, Leonard Clark, reports that he observed an anaconda swallowing a fully grown hog in the Peruvian rain forest (1953). We are reminded of his tale whenever we observe a factory trying to swallow a pig!)

Another way to keep the workload steady is to establish production routings based on product family. Often similar products require similar processing and utilize the same workers and equipment. If all products in a given product family follow the same routing, setup operations can be simplified and workers become familiar with the product. Process improvements for one product may be applicable to the entire product family. Common routings are a natural consequence of organizing the factory by product family (or value stream) rather than by functional departments.

We must make an important distinction between *workload* and *capacity*. Today's factory should strive for a steady workload and a balanced workflow. Today's factory should not strive for *balanced capacity*! Nevertheless, a popular approach to process improvement involves balancing the capacity by finding and improving the process that is constraining the overall production flow. In common language this is called *breaking the bottleneck*. This approach has led to many dramatic improvements in factory operations, but it has an inherent weakness: taken to its logical conclusion all processing stations eventually become "bottlenecks." If all stations are bottlenecks, then they all must be protected with excess WIP, which, in turn, causes increased congestion, longer cycle time, longer lead time, higher production cost, and the entire battery of problems we have discussed earlier.

## Reduce Variability Throughout the Value Stream

It turns out that variability reduction is one of the most powerful approaches to reducing lead time and improving factory performance in general. Virtually every one of the so-called JIT techniques is intended to reduce some form of variability (Crawford and Cox 1991). However, we emphasize again that it is dangerous to cherry pick from a list of favorite JIT practices (Inman 1993)! Based on our own factory experience, numerous case studies, and rigorous scientific research, *simply implementing JIT techniques does not produce any measurable improvement in factory performance* (Sakakibara, Flynn, Schroeder, and Norris 1997: 1256). Techniques are most valuable when they are part of a holistic management strategy that helps the factory reach its goals and objectives.

Remember that in this context the term *variability* refers to the *variability of processing time*. This variability comes from many internal and external sources such as irregular or inconsistent work procedures, planned and unplanned equipment downtime, long setup operations, and late deliveries from unreliable suppliers. Furthermore, management decisions such as purchase order size, production lot size, maintenance policies, equipment purchases, and shift schedules also contribute significantly to processing time variability.

There are three ways to handle variability: *eliminate it, reduce it, or adapt to it*. Our strong recommendation is to eliminate it wherever possible. However, due to the stochastic nature of the universe and today's highly uncertain manufacturing environment, it is more realistic to reduce it as much as possible. After variability has been reduced, the factory must find ways to adapt and respond to the remaining variability.

We use stockcar racing to illustrate different approaches to variability reduction. In a typical stockcar race, the *pole position* is won during the pre-race time trials. Usually a time trial lasts three laps, and the official *lap time* is determined by each driver's fastest lap. The car with the fastest lap time wins the pole position and has the privilege of starting the race in first place. If top speed were the only requirement for winning a race, stockcar racing would be pretty dull. The fastest car would start in first place, and the other cars would simply chase it throughout the race. However, the car with the highest top speed is not necessarily the car that wins the race. The car that drives to the winner's circle is the one that travels the total distance in the shortest amount of time. This is *average speed* as opposed to *top speed*.

What prevents a racecar from traveling at top speed for the entire race? Well, there are several reasons. The most obvious is the need for fuel and tires along the way. It is amusing to visualize how the pit crew might refuel the car and change all four tires while it is traveling at 200 mph around a track! Another reason the driver might slow down is to navigate around slower cars or to avoid an accident. If oil leaks onto the track or if an accident occurs, the cars must slow down to a safe speed or stop while the track is cleared. So, there are several good reasons for stopping or reducing speed during a race.

In this illustration we consider the pit stops. The driver must bring the racecar to a complete stop several times during a race to refuel, replace worn tires, have the windshield washed, and maybe even get a drink of water. There is no way to avoid refueling and changing tires, so successful race teams have learned how to reduce the amount of time required for each pit stop. The pit crew practices for hours each day, and the activity is choreographed with surprising precision. When the racecar comes into the pit for fuel, the team is waiting with two full cans of fuel and four fresh tires. The "gas man" is holding one can in his arms, and he is ready to jump over the wall and begin fueling the racecar. Similarly the "tire man" is holding two tires, one under each arm, and he is ready to jump over the wall the instant the car pulls into the pit. While the car is being fueled, other team members are changing tires, washing the windshield, and performing any other checks or adjustments that might be necessary. Since pit stops cannot be eliminated, the race team has learned to perform all these functions in parallel in order to reduce their negative effects as much as possible. Mathematically, this is equivalent to reducing the *variability in lap time* associated with the pit stops.

The stockcar race is analogous to a factory. In both cases, variability is sometimes unavoidable. Changing tooling, setting up equipment, loading a different kind of material into a machine, running out of production material, having a quality problem that leads to scrap or rework, and continual equipment failure are all common causes of factory variability. When variability *can* be eliminated, we should do so at once. When variability is unavoidable, we must learn to adapt to it and minimize its negative effects. For example, when production must be stopped because of a changeover operation, the variability in processing time can be minimized by:

- having tooling prepared in advance
- having material available at the worksite
- completing as much of the setup as possible while the machine is running
- having fixtures, pins, and slots instead of infinite adjustment
- having the necessary people on site.

As in the stockcar race, the critical variable in today's factory is time.

We recently presented this racecar analogy to a group of engineers. One engineer protested the validity of this analogy, declaring that racecar drivers are *in competition!* Aren't factories in competition too? Aren't the stakes much higher for today's factory than they will ever be in car racing?

Reducing cycle time is an important manufacturing strategy because it supports many worthwhile business goals such as faster delivery, greater flexibility, lower cost, higher revenues, and higher profits. Short cycle time means short lead time; lead time is directly related to cost and inversely related to total revenue (Duenyas and Hopp 1995: 49). Short lead time means low cost and high revenue, and long lead time means high cost and low revenue. This underscores the importance of lead time reduction for today's factory. Yet many factories still struggle with reducing lead time. The five strategies presented in this section are intended to help our readers develop their own effective approaches to reducing lead time. We have not suggested any particular tools or techniques but, instead, have offered some proven, practical guidelines for lead time reduction.

# Summary

A factory can be a sustainable source of competitive advantage, especially if the manufacturing strategy guiding the factory supports overall business goals. The alignment of manufacturing strategy with corporate strategy ensures that shop floor policies and managers' decisions support the company's broader goals and objectives. Proper alignment of the factory's goals and the corporate or business goals is essential to the success of today's factory.

Surprisingly, 90% of American manufacturing companies have no manufacturing strategy! It is the fortunate factory that has the benefit of well-aligned, clearly communicated, and widely understood long-term direction. Developing a manufacturing strategy requires first understanding the overall business objectives and the current state of the manufacturing operation. Only then can a strategy be developed that can guide the factory from where it is to where it needs to be.

There is a single manufacturing strategy that produces a number of very beneficial effects. This strategy is the *reduction of cycle time*. The benefits include short lead time, greater flexibility, less congestion, minimal investment in inventory (WIP), lower production cost, increased revenue, and ultimately higher profit. This chapter provides several clear guidelines for reducing cycle time: look for inventory, the "flower of all evil;" keep production material flowing toward completion; synchronize production; keep the workload steady; and reduce variability throughout the value stream.

Both lean manufacturing and factory physics highlight the effectiveness of reducing cycle time by reducing variability. In fact, virtually all "JIT techniques" reduce some form of manufacturing variability. Some researchers (Schoenberger 1986) refer to variability as the universal enemy and highlight the importance of continual improvement. The next chapter explains how today's factory can reduce variability and improve performance by achieving continual improvement and operational stability.

*Chapter 7*

# Operational Stability and Continual Improvement

## Introduction

This chapter introduces the concepts of operational stability and continual improvement and explains how they can be the foundation of a dynamic and responsive manufacturing system. Operational stability improves factory performance by reducing both internal and external sources of production variability throughout the value stream.

Recall the importance placed on variability reduction by both lean philosophy and factory physics. In Chapters 5 and 6 we emphasize that variability reduction is one of the most powerful approaches to reducing lead time and improving factory performance in general. The detrimental effects of variability are long cycle time, long lead time, high WIP, low throughput, and many other production problems. Low variability leads to short lead time, excellent customer service, low cost, and high profit! Therefore, variability should be reduced wherever possible.

Today's factory must improve continually to remain competitive in an ever changing and uncertain environment. Factories that expect and encourage change have the flexibility to adapt and improve their operations. However, action must be taken to ensure that improvements are incorporated into the daily routine and become a permanent part of the standardized procedure. Otherwise, the improvements will erode quickly and the benefits will be lost. Improvement without standardization cannot be sustained.

In this chapter we discuss kinds of wastefulness that are prevalent in manufacturing, but we point out that one company's waste is another company's competitive advantage. Eliminating waste is not a holistic management strategy. However, wastefulness in a manufacturing operation can be an indicator of other, more serious problems. By identifying and solving these problems, variability is reduced, and the operational stability of the factory is improved.

## Operational Stability

The term *stability* implies reliability and as little variability as possible. Stability, however, does not imply stasis. The term *operational* implies that every process in the production operation should be stable. So, the effect of operational stability is to minimize variability throughout the production stream.

There are many manufacturing practices that support the principle of operational stability. Because of their stabilizing effect, these practices are particularly important in today's factory. For example, standardized work, supplier involvement, productive maintenance, and robust product and process design all contribute significantly to operational stability. Customer management also contributes to operational stability since wildly fluctuating orders are a source of considerable variability. In short, we include as part of operational stability any activity that seeks to stabilize the manufacturing operation. Entire books have been written about individual elements of operational stability, but our purpose here is to describe briefly several of them and demonstrate how they contribute to the overall reduction of variability.

### *Standardized Work*

Standardized work ensures that production operations are performed the same way each time. It minimizes the natural variation in work performance from one operator to the next, and it helps any given operator perform the job consistently. Standardized work procedures are maintained and updated by the shop floor workers who actually do the work, although engineers often develop and write the original documents.

Standardized work procedures are displayed at the work site and followed enthusiastically. What compels the workers to follow a designated work procedure? Part of the answer lies in ownership: workers who participate in the development of the standardized work feel pride of ownership. There is also peer pressure associated with standardized work. Since the workers develop the standardized procedure in collaboration with one another, there is a certain amount of embarrassment if a worker performs the job another way. We believe the strongest reason for operators following the standardized work procedure is that the workers have agreed unanimously that it is the best way to do the work.

There are three key points to remember about standardized work. 1) Standardized work is standardized, not static. When a better way is discovered or when adjustments are required, the standardized work procedure is updated. 2) Standardized work supports production stability because the work is performed the same way each time. Defects, deviations, and discrepancies are recognized easily, and corrective action can be taken immediately. This reduces variability and enhances the stabilizing effects of standardized work. 3) Standardized work is essential for successful *kaizen* or continual improvement. *Kaizen*, a compound Japanese word meaning literally "change for the better," will be discussed in detail later in this chapter. It is surprising but true that standardization is an important prerequisite for permanent and cumulative change. This is primarily because standardization eliminates the "saw-tooth" effect (Figure 7.1).

Figure 7.1 shows that by using the typical improvement approach, gains are achieved but then partially lost. This is because the improvements are not formally adopted and are not included explicitly in a revised standardized work procedure. Failure to incorporate improvements into the standard allows and actually encourages reverting to the old way because the expectations are not changed. With standardization the improvements are permanent, so subsequent improvements begin at successively higher levels of performance. Standardization protects the improvements by revising the expectations at each new performance level. It should be noted that in both approaches (Fig. 7.1) the amount of effort required and the magnitude of each incremental improvement are equal. However,

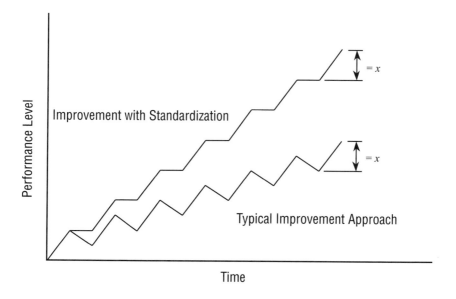

*Figure 7.1 Continual Improvement and the Saw-Tooth Effect*

a much greater long-term improvement is possible through standardization.

The saw-tooth effect is not a hypothetical phenomenon. In fact, it is present in the vast majority of process improvement programs. It stems from a reluctance to update standards and procedures after a better way has been discovered. If the expectations are not changed to reflect the revised methods, there is a strong tendency to revert to the old way. We observed an example of the saw-tooth effect in a small manufacturing company in southeastern Michigan.

This particular plant has approximately twenty screw machines and several other machines such as broaches, drills, mills, and lathes for secondary and finishing operations. The manager is evaluated based on a standard daily production rate established at 70% of the true machine production rate. One of the operators conjectured that by improving the tooling, the production output of his machine could increase by 50%. At this level the customer demand could be met by the output of one machine rather than using a second machine to supplement the first machine's output. The tooling was redesigned, purchased, and installed on the machine, and the expected increase in productivity was realized immediately. Figure 7.2a represents the daily output before and after the tooling improvement.

After the improvement was made, the standard production rate was not changed. The manager used the productivity increase to offset other jobs that did not meet their standard production rates. Before long, the operators realized that their expected daily production had not changed, and they began reducing their

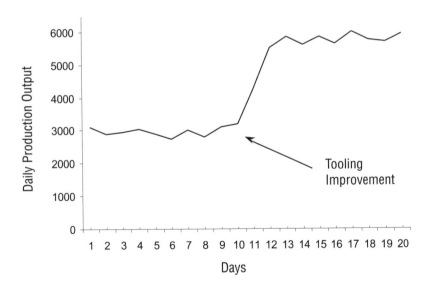

**Figure 7.2a**  *Daily Production Output Before and After Tooling Improvement*

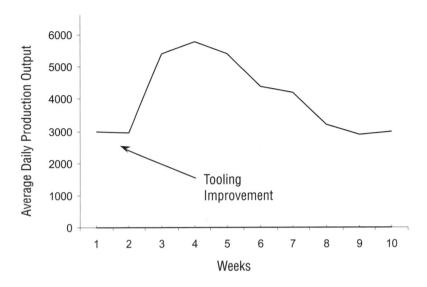

***Figure 7.2b*** *Average Daily Production Output Before and After Improvement*

output until they were once again producing at (or slightly below) the unchanged standard for the operation that had been "improved." Figure 7.2b is an extension of Figure 7.2a to include two months beyond the date when the new tooling was installed. The operator who had been instrumental in driving the project was disheartened and lost interest in participating in future improvement efforts. The manager lost his chance to profit from the increased productivity, and the company failed to realize any significant long-term benefits.

## *Supplier Involvement*

*Supplier involvement* is based on the philosophy that suppliers are partners for success rather than adversaries to be manipulated and conquered. In Chapters 4, 5, and 6 we emphasize the importance of short cycle time in running today's factory. This means that the time from procurement of raw material to delivery of finished goods to the customer should be reduced as much as possible. When suppliers are unreliable, deliver sporadically or in large batches, or when the delivered parts are of poor quality, the factory must hold additional inventory to avoid disrupting production. This has the negative effect of increasing cycle time and production cost. Supplier performance has a profound influence on factory performance, and it is advantageous to treat suppliers as partners for success.

It is important to recognize that the term "supplier" does not necessarily imply an external source of goods. It can also refer to divisions within a company or departments within a single factory. The customer-supplier relationship exists wherever a subsequent process uses output from a previous process.

Supplier quality certification is one important aspect of supplier involvement. This entails auditing the supplier's quality system to gain confidence that the supplier can produce and deliver material that meets the requirements. The supplier certifies that each delivery conforms to relevant specifications. Consequently, the material supplied is only of the highest quality. This aspect of operational stability reduces variability in three ways. First, the need for inspection at the receiving dock is eliminated. The supplied material can be delivered directly to the production line without waiting for inspection and disposition. This practice, sometimes called "ship-to-WIP," reduces significantly the total time that material spends in the factory. Second, the risk of parts shortages due to poor quality is eliminated. Third, the supplier can deliver in much smaller quantities since the supplied material is no longer inspected. In traditional quality assurance systems, each lot received from a supplier is inspected to ensure that the material meets specifications. This can be extremely time-consuming and labor-intensive. Many companies use this reasoning to justify receiving large lots from their suppliers.

Supplier involvement is much more than just high quality. If the factory and its suppliers are *partners for success*, then the factory must be willing to share information about cost, inventory, and production. By sharing cost information the factory and the suppliers can arrive at pricing structures and delivery schedules that are mutually fair and profitable. Shared inventory information ensures that the supplier doesn't produce and deliver material that is not needed; it also ensures that the factory does not stock-out or run short of required material. Sharing actual production requirements eliminates surprise orders and wildly fluctuating delivery demands from the factory. Some manufacturers even post their inventory and production data on the Internet so suppliers have real-time information on which to base their own production schedules.

Another aspect of the supplier relationship is product design involvement. This allows the supplier to influence the design so products can be manufactured efficiently. If the supplier is involved in the product design phase, it is more likely that the supplier will be able to manufacture the product easily. It is far better to design a product that is easy to make than to design a complicated process to accommodate a needlessly difficult product. The ease of manufacturing can be improved dramatically by minor and inexpensive improvements in initial product design. Modifications to the manufacturing process after the product has already been designed are costly and have relatively little effect on the ease of manufacturing.

In Chapter 5 we saw that variability in a production routing is influenced dramatically by transfer batch size. Similarly, variability in the supply chain is influenced dramatically by the size of shipments from suppliers. This means that supplies should be delivered in small lots as frequently as needed. The supplier will also benefit from making regular, small lot deliveries since it is much easier to fill regular, frequent orders than to fill sporadic, unexpected, "monster" orders. Also, by delivering small lots frequently, the supplier has the opportunity to reduce its own production variability, thereby improving its cost, quality, and dependability.

Purchasing managers are often reluctant to contact suppliers to request more frequent delivery of smaller lots. This hesitation seems to be based on two assumptions: that the supplier will flatly refuse and that delivery costs will increase. In one case when we encouraged a purchasing manager to call several suppliers, she was flabbergasted when a supplier replied that many customers were making the same request, and the supplier expressed surprise that she had waited so long to make an inquiry! At a small spring manufacturer where we have consulted, the raw material was purchased in 40,000 pound shipments. This was to ensure a low price through volume discounts. When the owner asked about receiving smaller shipments weekly according to pull signals (Chapter 9), he was given full cooperation and a significant price decrease!

Ironically, many factory managers resist frequent deliveries because they are considered less "efficient." Frequent deliveries are most definitely less efficient for individuals and departments such as receiving, purchasing, and material handling. What is essential to keep in mind is whose efficiency takes precedence. Very often people and departments use "efficiency" to mean their own efficiency. They fail to consider activities beyond their own task or area and do not evaluate the broader production process.

In the case of smaller, more frequent deliveries, the purchasing department places releases more often, and the receiving department accepts supplies more often. More effort is expended, and, consequently, these departments are "less efficient." However, the variability of the cycle time is reduced considerably when smaller lots are accepted. This means there is less material to move, count, stage, and store; the material is less likely to become damaged or obsolete; and there is less investment in inventory. The factory is more flexible and can respond more quickly to changing customer requirements. The "efficiency" that is most important is the efficiency of the overall manufacturing enterprise. This is the goal of today's factory. Toward this end, *the benefits of regular, small lot deliveries cannot be overstated.*

The payoff of supplier involvement is that components of certified quality that were designed in collaboration with suppliers are delivered frequently in small lots at a fair price just in time to accommodate production requirements. Relationships like this can only be built through communication and mutual involvement. Supplier involvement is clearly a huge advantage to today's factory and a key aspect of operational stability.

## *Productive Maintenance*

Operational stability is influenced directly by the condition and availability of production equipment. The purpose of productive maintenance is to sustain the normal operating condition of machinery and tools and to eliminate all machine breakdowns. Obviously, machine downtime can never be eliminated completely, but *unexpected downtime should be reduced to zero.* Unexpected downtime occurs when equipment breaks down suddenly during a production run. This contributes variability to the production operation, lengthens the cycle time, increases the customer lead time, and increases cost!

*The goal of productive maintenance is 100% equipment availability.* By this we mean that production equipment is available for use at any time. This is not the same as 100% machine utilization, and the distinction is an important one. To illustrate this point we often use the analogy of a personal car. We want to be able to get in our car and drive where we need to go any time of the day or night. We do not necessarily want to drive the car 24 hours a day. (The issue of machine utilization will be addressed extensively in Chapter 12.)

Productive maintenance is a shop floor activity, and it takes place at the machine. The operator's involvement is crucial. After all, the person who uses the machine regularly is most likely to notice an abnormality. The operator should be encouraged to perform daily equipment inspection. This provides the opportunity to correct minor problems that might otherwise be overlooked. In practice this may not be as simple as it sounds. The operator may not have enough technical knowledge about the equipment to diagnosis the problem. The operator may not have a forum for communicating concerns, or the concern may be dismissed as a complaint. There may be no one to respond, or the concern may be given low priority by the maintenance department and nothing is done.

We have found that operator check sheets are an excellent way to manage the daily maintenance of equipment and to report its condition. Items that need to be checked, cleaned, lubricated, or adjusted daily or periodically are listed. An operator initials when a maintenance procedure has been performed, and space is provided to note abnormalities requiring the maintenance department's attention. Maintenance personnel, supervisors, engineers, and managers review the check sheets daily, and action is taken immediately when problems are discovered. This reinforces the use of the check sheets and the value of operator involvement. If a problem cannot be fixed quickly, the operator needs to be informed what the problem is, what is being done to resolve it, and approximately how long it will take. Again, this shows respect for the operator and reinforces involvement and commitment.

Another important aspect of productive maintenance is cleaning, and, in fact, cleaning can be viewed as an inspection process. Let's consider the car again. The car is dirty after being driven during a long Michigan winter when roads are covered with salt and sand. While washing the car by hand, the owner is likely to notice dings in the paint, chips in the windshield, or loose trim. It is highly likely that the owner who drives the car every day will see blemishes that had gone unnoticed for weeks or even months. The close scrutiny that goes with hand washing allows the owner to observe these "abnormalities."

By analogy, the act of cleaning production equipment by hand gives operators an opportunity to observe abnormalities such as loose fasteners, missing components, misplaced tooling, or improper settings. (Suzaki (1987: 119) highlights the importance of tightening fasteners with an anecdote about a company that reduced machine breakdowns by 80% just by tightening tens of thousands of bolts in the plant.) Also, cleaning by hand has the psychological benefit of nurturing the operator's sense of ownership in the production process. Productive maintenance results in greatly reduced variability through capable employees taking responsibility for their own equipment.

### *Robust Process Design*

One good way to stabilize a production process is to use standardized work procedures. If every operator performs the process the same way, variability decreases naturally. There are, however, ways to augment the benefits of standardized work, and we will mention a few of these here. Standardized gauges and fixtures are important tools for stability, and they ensure that each workpiece conforms to specifications. This was crucial for interchangeability in the early days of mass production (Chapter 2), and it is equally important today. Another way to ensure process stability is to use simple, single-purpose machinery. Simple machines are far less expensive to build, purchase, and maintain, and they can be incorporated easily into a sequential, product-oriented layout.

It is important that engineers design processes that are *robust* and *insensitive* to minor variations in the input parameters. In a nonrobust process, varying the input parameters causes the characteristics of the finished product to vary. In a robust process, varying the input parameters does not cause the characteristics of the finished product to vary. A robust process, therefore, is one that can tolerate variation in the input parameters without the output of the process being affected. Figure 7.3 is a graphical representation of robust and nonrobust processes.

A factory can be a rough place, and production processes must be able to withstand heavy use and a certain amount of misuse or abuse. Sensitive and delicate laboratory devices may not endure the factory environment, so production equipment must be durable and the processes robust.

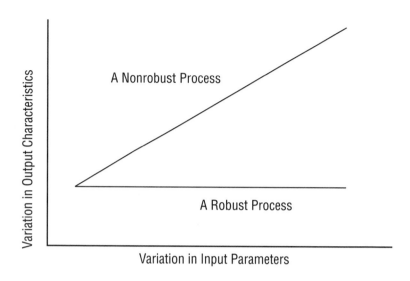

*Figure 7.3 Robust and Nonrobust Processes*

We are reminded of a company that tried to save money during a machine-building project. This company wanted to construct a high-spindle-speed cross-drilling machine. The engineer in charge of the project purchased some used skateboard bearings at a garage sale and installed the bearings in the new machine. When the machine was turned on, it made so much noise that it could be heard throughout the factory, in the corporate offices, in the parking lot, and all the way to the street! The holes generated by the cross-drilling machine were badly out-of-round and oversized. Fortunately for everyone the bearings lasted only a few hours, and nobody was hurt. The worn skateboard bearings were replaced by high-speed spindle bearings, and the machine has quietly drilled millions of precision holes since then.

## Robust Product Design

*Robust product design* refers to products that are designed for ease of manufacture. Tolerances are based on engineering principles and functional requirements rather than being arbitrary, needlessly restrictive, or chosen out of fear or uncertainty. For example, functional gauging verifies the fitness for use of a workpiece without being overly restrictive on the noncritical dimensions. Geometric Dimensioning and Tolerancing (GD&T)—a tolerancing system based on product functionality—is highly recommended. GD&T takes into account the intended use of the workpiece and provides the widest possible latitude for dimensions. Robust products are designed so minor variations in the manufacturing process can be tolerated without the product being rendered unacceptable or unusable.

Ideally, many functional groups in a manufacturing organization contribute to robust product design. Marketing and customer service departments best understand the needs of the customer. The production group knows the manufacturing process more comprehensively than other departments. The product engineers can modify the product to accommodate the manufacturing process while still meeting the customer's requirements. The designers take into account the manufacturing process and the quality assurance requirements. The purchasing department is familiar with the availability and scheduling of raw materials, components, and outside services such as plating or heat-treating. The quality department understands the needs of the customer, the product design, and the manufacturing process. The finance department best understands the company's investment strategy and can assist in equipment and process selection. We can see that robust product design requires the input and involvement of nearly every functional group in the organization.

## Continual Improvement

*Continual improvement* refers to changes that are gradual and incremental as well as changes of a more dramatic nature. (The Japanese term for continual improvement, *kaizen*, actually means "change for the better" although it has acquired the connotation of extreme and radical change (the familiar *kaizen* event).)

There are two levels at which continual improvement takes place in a factory: the process level and the system level. Process improvement is detailed and narrow in scope, and system improvement is broad and holistic in scope. Both are necessary if the factory is to benefit from improvement efforts. Process improvement alone cannot produce system-wide advantages, and system improvement requires that specific processes within the system be modified.

Process improvement is continual and incremental. It takes place on the shop floor with the people who are involved in the specific process. In practice, shop floor workers can improve their own processes without the assistance or involvement of engineers or managers, but there is no reason to exclude anyone who wishes to be involved in the improvement effort.

Process improvement focuses on the details and methods of a specific production process. It can include the placement of tools, presentation of material, or arrangement of machines in a work cell. Fixtures or devices that automatically load or unload machines are often installed as a result of process improvement. Process improvement involves scrutinizing how people get their work done, and its goal is to minimize the time needed for material to flow through the area. Activities that do not add value to the product, such as unnecessary walking to get supplies or looking for tools, are reduced or eliminated. Of course, as with any change, the standardized work procedure should be revised.

Improvement should not be a special event that takes place sporadically. It should be an ongoing part of the workday. To make sure factory improvement is continual, we recommend a cyclical approach. Figure 7.4 represents a typical improvement cycle. First a problem or area that needs improvement is selected. Then the root cause of the problem is identified and a possible solution is devised. The solution is implemented and tested for effectiveness. If it works, the process is standardized to reflect the new method. If the solution is not effective, the "cause" that was identified is reexamined to make sure that it was truly the source of the problem. Once the true source of the problem has been verified, a new solution is developed, implemented, and tested. After the solution's effectiveness has been established, the standardized work procedures are updated, and the next problem or area needing improvement is addressed.

Finding the source of the problem is a serious endeavor. It is more than simply observing the symptoms; it must be a true root cause determination. One popular technique for doing this is "the five why's." Using this technique, the question "why" is asked repeatedly. Each response is interpreted as a symptom

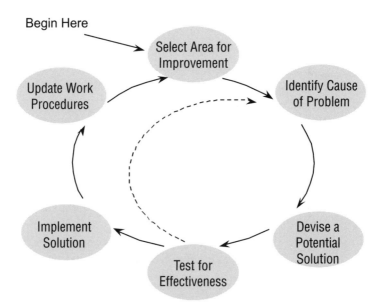

*Figure 7.4 A Factory Improvement Cycle*

or a superficial cause, so the question is asked again. Layers of "reasons" are stripped away like layers of an onion until finally a core reason or root cause is identified.

Recently we used this technique to help solve a problem we encountered in a factory. This particular manufacturing company reorganized its assembly area into several single-piece flow work cells. Each cell has a production display board on which the workers record important information such as the number of assemblies completed and any problems encountered. When management reviews the charts, the "problems" listed are often superficial, but they provide clues for finding the underlying root cause. One afternoon the production rate of a work cell fell far below normal. The team wrote on the chart that the "button building was taking too long." By repeatedly asking why, the real cause of the problem was eventually uncovered.

**Assembly area:**

| | |
|---|---|
| 1. Button building is too slow today. | Why? |
| 2. We have to crimp the buttons. | Why? |
| 3. Buttons are "bad." | Why? |
| 4. The actuators are cockeyed. | Why? |
| 5. The holes do not line up properly. | Why? |

**Fabrication area:**

| | |
|---|---|
| 1. The actuators came from the supplier that way. | Why? |
| 2. We accepted them because they fit our gauge. | Why? |
| 3. They were "to print." | Why? |
| 4. The engineering drawing allows the misalignment. | Why? |
| 5. Traditional tolerances were used rather than GD&T. | |

So, in this case, the slowdown in assembly was not caused by lack of vigilance on the part of the button builder. The engineering drawing allowed misalignment because of a stack-up of tolerances. When the supplier made a new "tool," the tolerance stack-up allowed the misalignment problem to occur.

Whatever problem solving methodology is used, it is important to find the true cause so an effective solution can be developed. Often excellent solutions are developed and applied to superficial problems without ever addressing the root cause. In the example above, the crimping of the actuators could have been considered "the solution," and *the true cause would never have been addressed*. Once the root cause has been determined, a solution should be developed and applied so the problem never occurs again.

There are many ways to develop solutions. Often group problem solving techniques are effective. This has the advantage of involving several people who are familiar with the product or process. People who are close to the problem have a vested interest in finding a permanent solution.

We do offer a caution about group problem solving: groups of people do not inherently develop better solutions than individuals do. It is even possible for the group solution to be worse than the worst individual solution. On the other hand, it is also possible for the group solution to be better than the best individual solution. To achieve the desired synergy, the group should be the proper size, comprise the right people, and use proven problem solving techniques.

## System Improvement

System improvement is the big-picture approach. It focuses on macroscopic, system-wide changes and the efficiency of the entire production system. System improvement seeks to create a smooth flow of material throughout the value stream. This is done by reducing variability (Chapter 5) and eliminating wasted effort wherever possible.

What constitutes wasted effort? What is considered wasteful in a manufacturing environment? Different companies have different answers to this question and different ways of categorizing waste. For example, the Ford Production System Institute prominently displays six manufacturing wastes. Canon identifies nine wastes, and Toyota officially recognizes seven categories of waste. Table 7.1 lists the categories of waste identified by these companies (Nicholas 1998: 75, 80-81).

| Canon | Toyota | Ford |
|---|---|---|
| WIP | Inventory | Inventory |
| Defects | Defects | Defects |
| Equipment | Processing | Processing |
| Expense | Motion | Motion |
| Indirect labor | Waiting time | Waiting time |
| Human resources | Transportation | Transportation |
| Operations | Overproduction | |
| Planning | | |
| Start-up | | |

*Table 7.1 Common Categories of Manufacturing Waste*

Listed below are the wastes that we feel are most prevalent in manufacturing. The first seven are similar to the ones Toyota recognizes. We have added an eighth equally significant category of waste: the waste of human potential. Our inclusion of this eighth waste is completely compatible with the science of factory physics and the philosophy of lean manufacturing with its emphasis on respect for humans. Wasted human potential is much more than just idle or underutilized workers. Individuals can make creative contributions to improve and adapt manufacturing processes, and the people working in a factory represent a competitive advantage that too frequently is overlooked. There is nothing magical about these eight categories. They simply help us take a hard look at the entire stream of production and somehow separate the activities that create value in the eyes of the customer from those activities that do not.

1. Overproduction
2. Inventory
3. Defects
4. Processing
5. Motion
6. Waiting time
7. Transportation
8. Unrealized human potential

**Overproduction**

What is overproduction? *Overproduction* is producing more, sooner, or faster than required by the next process or by the customer. Producing *more* than required is the most obvious form of overproduction. The excess material must be moved, counted, staged, stored, inspected, reworked, and so on. Producing *sooner* causes the same problems as producing more, although these problems may be short-term. Producing *faster* also results in material accumulating somewhere in the production process. The excess material resulting from overproduction lengthens cycle time and lead time, and it increases production costs (Chapter 5).

A nonmanufacturing analogy to overproduction involves air traffic control. Many of us have experienced the delayed departure of a flight because of weather conditions *at the destination* that prevent the airplane's landing. There are good reasons for this policy. If the flight departs on time, it is forced to fly in a holding pattern until it can land safely. This wastes fuel, fatigues the pilots and crew, puts needless wear and tear on the aircraft, and it can overtax the air traffic control staff resulting in a dangerous situation. The flight does not take off if it can't land safely, and it doesn't arrive at its destination until the airport is ready. This policy seems very reasonable when the subject is air traffic control. Producing goods before the next process is ready for them is *exactly analogous* to the airplane departing without being able to land safely at its destination.

Another nonmanufacturing analogy involves a well-tuned automobile. In this analogy a traveler is driving from Miami, Florida, to Nashville, Tennessee, to see an important client. When the traveler reaches Nashville, he realizes that he is making excellent time. The car is running well and is getting great mileage. The driver decides to take advantage of his good fortune, and he continues driving, eventually reaching Spokane, Washington. Unfortunately, his client is still back in Nashville. In this case the "efficiency" of the car only takes the traveler farther from his goal. Moreover, trying to take advantage of the well-running car has associated costs, namely fuel, operator time, and wear on the automobile. So, by analogy, extending a production run of an operation serves no purpose if the excess product is not needed.

Overproduction is considered the worst waste of all because it leads to nearly all the other wastes. Overproduction causes excess inventory, and it can hide defects and hinder the search for their root causes. Overproduction can create a need for additional processing, such as rust protection or rework. Overproduction also leads to more conveyance effort because the excess material must be moved, stored, and otherwise handled many times. Waiting time of the material, the largest component of cycle time, is increased by overproduction since the material is produced before it is needed.

Overproduction is also disrespectful to people. When operators are expected to work hard producing quality parts that aren't needed for many weeks or months, motivation is low. Furthermore, if the parts produced are defective, operators can view their work as wasteful or even detrimental to the company. We recently observed an operator making subassemblies for a high-precision product. The subassemblies must be made with exactness and care. In the final step of the subassembly operation, the component is sealed with epoxy. On this particular day, the operator stopped us to show us how productive and diligent she had been, producing several weeks' worth of subassemblies! The next day we saw the same operator chipping the subassemblies apart because one of the components was incorrect. She was terribly disheartened! If she had not been expected to "overproduce," she would have had only a few subassemblies to repair. Overproduction decouples those who build the product from those who use it, and it contradicts the concept of the next process being a customer.

### Inventory

Inventory is not "the root of all evil" as some advocates suggest (Suzaki 1987: 17), but *excess* inventory is a tremendous waste. Inventory is often used to insulate the factory from problems, such as long setup operations, poor maintenance practices, unreliable suppliers, and improper production control policies (Inman 1993: 41). Therefore, it can be viewed as an indicator of problems rather than as a problem in itself. The capital tied up in excess inventory can be considerable, but the "cost" also includes the wasted effort of moving, tracking, storing, and otherwise "managing" the unneeded material. Excess inventory is also subject to damage or obsolescence. These costs can be much higher than the actual material costs.

## Defects

Defects waste time: time to detect, time to repair or replace, time to sort, and the time it takes to produce the defective products in the first place. Defects also waste material and create scrap. Defects waste human potential and worker effort. Furthermore, from a factory physics point of view, defects directly contribute to high variability. (The *scv* of effective processing time (Chapter 5) increases with an increase in defect rate.) This in turn causes more congestion, longer cycle time, longer lead times, higher WIP, higher production costs, and the entire battery of problems related to high variability!

## Overprocessing, Conveyance, and Unnecessary Human Motion

Overprocessing is another waste. It entails doing more than required minimally to transform material into an acceptable product. The waste of conveyance refers to excessive material movement, including transportation, stacking, and staging. This waste is distinct from the waste of unnecessary human motion. Unnecessary human motion is a waste of time and energy; it is tiring, stressful, and disrespectful to the workers.

## Waiting Time

Making people wait to do their work is also disrespectful and wasteful. We find that most workers prefer to be occupied, and they find it aggravating and stressful to wait for parts, maintenance repairs, or changeovers. There is also considerable fear associated with waiting time. This includes fear that people might be sent home or laid off, fear that the waiting delay is a sign that the company is struggling financially, or fear that the supervisor is mismanaging the department. Making people wait has serious psychological implications.

There is also a waste associated with material-waiting time. Remember, if the goal is to reduce cycle time, and if the largest component of cycle time is waiting time (Chapter 4), then material-waiting time should be reduced wherever possible.

## Human Potential

The eighth waste, the waste of unrealized human potential, disregards the most significant asset of a factory: the cumulative wealth of employee experience and creativity. Human beings have the ability to solve problems, create new products, and improve processes. They can learn, invent, adapt, and teach. Furthermore, our employees are also the only assets that appreciate. In fact, they are one of our best competitive advantages: they cannot be duplicated by our competition.

## *Weakness of the "Eliminate Waste" Approach*

In the popular literature, "waste" is characterized as the universal enemy. Slogans like "eliminate waste" are becoming the *mantra* of lean manufacturing. Is eliminating waste an effective strategy for today's factory? Our answer to this question is based soundly in scientific reasoning and years of manufacturing experience: No! If eliminating waste is the central theme of an improvement effort, the benefits will be superficial. Part of the reason lies in the ambiguity of the term. First, the term "waste" has come to mean anything that is bad in the factory. While eliminating bad things from the factory is certainly beneficial, it is hardly a manufacturing strategy (Chapter 6). Second, one company's waste may be another company's competitive advantage (Chapter 1). For example, one company may hold finished goods inventory (waste) in order to respond more quickly to customer demand (competitive advantage).

There is another very significant reason why eliminating waste is not a good manufacturing strategy: variability (Chapter 5). When variability is present, the factory must buffer itself in at least one of three ways: excess inventory (waste), excess capacity (waste), or excess lead time (waste). If all the "waste" is eliminated from the system, the factory will be incapable of handling any variability.

Running today's factory involves more than chanting a *mantra* and stamping out waste. *Running today's factory requires understanding the strategic goals of the company and developing a manufacturing strategy that supports those goals.*

System improvement focuses on improving how material and information flow through the plant. Its objective is to minimize the cycle time. This often results in major changes to the physical layout of the plant and to the production control procedures. Often entire departments are incorporated into other areas of the factory. For example, a centralized welding department might be replaced by several small welding stations, each positioned at the appropriate location in the production sequence. System improvement usually results in significant, measurable performance benefits including shorter customer lead times and increased efficiency, productivity, and profitability. As with any improvement effort, changes resulting from system improvement must be incorporated into the operating standards and procedures.

## *Kaizen*

There is another approach to improvement that is used in some American factories. It is known by various names such as *kaizen blitz* (Day), *kamikaze kaizen* (Womack), and *drive-by kaizen* (Rother and Shook). We refer to this type of improvement activity as *paintball kaizen*. The paintball *kaizen* game plan is analogous to someone painting the living room walls by sitting on the couch and taking pot shots with a paintball gun. After the first few paintballs hit the wall, the living room just looks like a big mess. The color change is underway, but there is no discernable pattern, and the old color probably looks much better. Some areas are hit more than once; other areas remain the old color or, at best, are splattered with the new color paint. Eventually, this method can result in a complete color change of the walls, but only after a lot of wasted paint, several arguments with your mate, and maybe even a casualty or two. This approach to improvement is characterized by isolated local improvements that are not part of a holistic plan, although there may be an overarching goal.

When paintball *kaizen* is used in a factory, a particular process or area of a plant is chosen, and it is transformed by applying various lean manufacturing techniques. Often a team of outside experts leads these "*kaizen* events." Some companies even sell tickets to outsiders who want to observe, participate, and learn about *kaizen* activities. These events often lead to isolated islands of excellence, with no way for material to flow smoothly from one area to the next. For example, perhaps the appearance of a work area improves dramatically. Machines may be cleaned, lines may be painted on the floor, and unnecessary equipment and material may be removed from the area. Perhaps the setup time for a machine is reduced significantly. These are good and positive achievements, but if they don't improve the flow of material, then the benefit to the factory will be minimal. This improvement approach often alienates workers in the area who may or may not have been involved in the *kaizen* event. Also, workers resent outsiders changing work procedures without truly understanding the process. Improvement efforts that are not part of a long-term, system-wide strategy can actually hinder the operation of a factory.

# Value-Stream Mapping and Production Flow Diagrams

Value-stream mapping and production flow diagrams are "big picture" tools for improving a manufacturing system. Value-stream mapping is a technique for analyzing material and information flow. It can be used in a specific area of a factory or extended across company boundaries to analyze the entire value stream from raw material to finished goods. It is particularly useful in helping people "learn to see" the flow of material in their own factories (Rother and Shook 1998).

Value-stream mapping is essentially a flow-charting technique that uses icons to represent activities and locations. Variously styled arrows indicate the flow of information and material. Areas where material waits are identified, and the amount of material is represented in terms of quantity and time. This highlights the importance of cycle time in running today's factory. Manufacturing processes are also shown on the value-stream map, and relevant information such as processing rate, amount of material, scrap rate, downtime, and setup time is indicated. Value-stream mapping is an important analytical technique, but we caution our readers to avoid treating it as an end in itself.

Production flow diagrams distinguish between value-added and nonvalue-added activities in the value stream. The production flow diagram can take several forms. Photographs of each process can be mounted in sequential order on a display board. A video can be made showing each manufacturing process in the production of an item. The flow of material through a plant can be indicated by a bold line drawn on a plant layout with each process clearly labeled. The plant layout version of the process flow diagram is sometimes called a "spaghetti" diagram because the result usually looks like a plate full of *al dente linguini*. Most factory managers are stunned by the amount of nonvalue-added activity exposed in their operations (moving, counting, staging, inspecting, storing, receiving, issuing, reworking, and testing). This tool helps identify waste in the production stream, but it doesn't provide any direct help in reducing the cycle time in the factory.

## Summary

Pursuing operational stability is an effective way to reduce variability in the factory. It is also essential to successful improvement. "Stability" does not mean maintaining the *status quo* since today's factory must improve continually to remain competitive. Operational stability includes practices of standardized work, supplier involvement, productive maintenance, robust process and product design, and any other initiatives that seek to stabilize the daily operations of the factory.

Continual improvement takes place at two levels: 1) the detailed level of each process on the shop floor, and 2) the system-wide level of an entire manufacturing enterprise. Successful improvement depends on identifying and solving root causes rather than symptoms. A systematic approach to problem solving, such as the "5 whys" approach, can help us peel away layers of symptoms until the root cause is revealed. Solving superficial symptoms usurps valuable time and resources and does not result in the desired effect: the permanent elimination of the problem. Improvements that are incorporated into the operating standards are more likely to be used on a daily basis and to become a permanent part of the production process.

Continual improvement focuses on streamlining the flow of material through the value stream. Eliminating waste is an important part of factory improvement, but on its own it is insufficient. Isolated improvements cannot provide holistic benefit to the factory. Improving today's factory requires developing a manufacturing strategy that supports the company goals and implementing that strategy according to a well-conceived improvement plan. Operational stability and continual improvement are part of a manufacturing strategy that gives today's factory a strong competitive edge.

*Chapter 8*

# Just-In-Time Production

## Introduction

Just-in-time (JIT) is by far the most familiar and most widely publicized lean manufacturing technique. American manufacturing literature is replete with books and articles on this subject written over the past 20 years. Curiously, just-in-time remains one of the most misunderstood and misapplied techniques in American manufacturing. To some people, JIT means *Just Implement Techniques*, in other words, use "best practices" or any off-the-shelf technique that might solve a problem. To others, it is a system of integrated techniques that generates the promised benefits. To still others, it is a production philosophy; and to some it has come to mean anything that is good about manufacturing. We see "just-in-time" as a principle of manufacturing that is essential for today's factory.

*Just-in-time* means having the right part at the right place in the right amount at the right time. It is much more than a delivery technique; it is a core principle of lean manufacturing philosophy. JIT is an approach to production and operations management that seeks to streamline the flow of material and information. It results in far less variability and much shorter cycle times than other approaches to materials management.

When a manufacturing organization adopts just-in-time, inventory levels usually drop, often by a factor of ten. Yet, we emphasize that JIT is not a zero inventory policy. Reducing inventory *per se* is not an explicit goal of JIT; it is a consequence of reducing variability. Reduced variability and shorter cycle times are the reasons why JIT works; inventory reduction is a beneficial consequence.

Almost every element of JIT has as its purpose the reduction of variability. High variability leads to high WIP and long cycle times. Low variability leads to low WIP and short cycle times. Consequently, reducing variability is an excellent goal for today's factory.

The factory's flexibility to make what the customer wants when the customer wants it is increasingly important in today's economic environment. Flexibility is a powerful competitive advantage. Researchers from MIT and the General Motors Research and Development Center conclude that "increasing manufacturing flexibility is a key strategy for improving market responsiveness in the face of uncertain future product demand" (Jordan and Graves 1995: 577). JIT can provide the factory with that flexibility.

This chapter addresses several of the more important practices associated with just-in-time production. We focus on the scientific principles and strategic intent behind JIT rather than instructions for putting JIT into place. Understanding the principles and true intent should enable the reader to apply the practices in any manufacturing situation.

Implementing JIT is not easy, and transforming a factory requires a long-term focus and a devoted management team. The payoff is improved responsiveness to customers, much shorter cycle time and lead time, more efficient utilization of resources, greatly reduced inventory, less congestion, more flexibility, lower production cost, and higher profits. The elements of JIT that we have chosen are illustrative of a critical thinking approach to running today's factory.

## Level, Mixed Production

One essential practice in just-in-time manufacturing is level, mixed production or *heijunka*. It is a tool for scheduling production quantity and product mix based on customer demand. As a result, the factory produces *at the rate of customer demand*, making what the customer wants when the customer wants it.

In practice, level mixed production is easy to apply. For a given time period, such as a day or a week, all the orders for one product are combined and distributed evenly through the production schedule. Then the orders for another product are combined and also spread evenly through the production schedule. This continues until all products requested by customers are included in the daily production schedule. The result is a production schedule that is thoroughly mixed. (Mathematically, this is equivalent to reducing the variability in the production schedule.) Practically speaking, the demands on suppliers and on the production operation are also evenly distributed, and variability in demand for materials, equipment, and effort is minimized.

An aerospace components manufacturer with which we are familiar makes hundreds of different products. In any given week, the factory produces approximately a few dozen of these. Each product has its own production problems. Some products are far more difficult than others and require more time to produce.

In the past, this factory consolidated customer orders by product type and then released the orders to the shop floor. The effect was that huge orders were created. Filling a huge order in one production run depleted departmental supplies at an unexpected and accelerated rate, and greatly overloaded the production resources (equipment and people). Frequently production material ran short and some processes could not keep up with the sporadic surges in demand. Often customer orders could not be completed because partially assembled items were stacked and set aside until parts or equipment became available. When large orders were released, the factory became congested with WIP. Subsequent processing was delayed, and the effects propagated downstream throughout the production line.

The effects of uneven scheduling also propagated upstream. Processing "monster" orders sent sporadic peaks of high demand to the stock room and to upstream processes such as molding, welding, and stamping. Those areas were likely to deplete their materials in one fell swoop, forcing these areas to send "monster" orders to outside suppliers. The huge orders greatly increased the variability of processing time, which in turn increased cycle time, lead time, congestion, production cost, WIP levels, and stress levels of all people involved.

This company is now releasing smaller orders more frequently throughout the week, and the product type varies from order to order. The supplies are consumed at a steady rate and can be replenished more easily by the stock room and fabrication departments. The goal is to have all customer orders for a given product and time period blended together and distributed evenly throughout the production schedule. *Today's factory should process material in a sequence that has as little variability as possible.*

To further clarify this concept, we use the analogy of a backyard cookout where the chef of the day is grilling hot dogs and hamburgers. One logical approach to satisfying the hungry picnickers is to grill all the hamburgers first and then cook the hot dogs. Clearly this approach is more "efficient" and less stressful for the chef; the grill is loaded with one type of meat at a time, the cooking time is uniform, and only one cooking utensil is used at a time. How well does this grilling approach satisfy the picnickers? Well, the hot dog customers have to wait until all the hamburgers are cooked before they can have a hot dog, and the hamburger customers wanting seconds have to settle for a burger cooked in the first batch.

While we were writing this chapter, we were invited to a friend's home for a backyard barbecue. On our drive there we reflected on our barbecue analogy and wondered how the meat would be grilled. To our great surprise our friend cooked the meat in large batches, burgers before dogs. Dale had to wait for her dog, and

**Figure 8.1** *Waiting for a Hot Dog*

Charles had a chilly second burger!

Many backyard barbecue chefs have a different approach. They regulate what they put on the grill according to what the people at the picnic are eating. It is common to see hamburgers, hot dogs, and even an occasional *bratwurst* sizzling together on the same backyard grill. This may be less "efficient" for the chef, who now must load and turn more than one type of meat, each with its own utensil and cooking requirements. The benefit, however, is that the hungry picnickers, the "customers," are much happier. They get what they want when they want it, and the price is right too.

Even though level, sequential flow (*heijunka*) is a common practice in backyards everywhere, it has not gained popularity in our factories. In a manufacturing environment the economies-of-scale paradigm is deeply ingrained, and the idea of producing concurrently a variety of items is usually resisted. The acceptance and use of level, sequential flow requires an open mind toward mixed-model production or, at the very least, toward small-lot production with frequent changeovers (Chapter 11).

Some companies are experimenting with another way of mixing their production. This is known as *in-line sequencing* or, in the automobile industry, *in-line vehicle sequencing*. In-line sequencing resembles level, sequential flow because it also involves mixed-model production. However, the underlying principle is very different. Several days in advance, suppliers receive a schedule of exactly what the factory will make and the exact sequence of production. The suppliers send the necessary materials "just in time" to meet production requirements. In theory this works well. In practice, though, it has one major drawback: the system is dangerously inflexible. Once the schedule is set, it is virtually impossible to change it. This inflexibility magnifies the impact of supplier tardiness, parts shortages, equipment downtime, and any production line difficulties.

Returning to the barbecue analogy, in-line sequencing would require that someone poll all the picnickers to find out whether they want a hot dog or a hamburger, how many they want, and when they expect to be served. The in-line sequencing approach would result in each person having a hot, fresh meal prepared "just in time." What the picnickers wouldn't have is flexibility: the leeway to play badminton for an extra five minutes, to throw the Frisbee® to the dog one more time, or to change their minds about what to eat. If the picnickers changed what they wanted or when they wanted it, or if the chef cooked the "wrong" meat even once, all subsequent orders would be out of sequence.

In a manufacturing environment, the customer requirements change radically and frequently, and production seldom goes according to schedule. In most manufacturing companies, the uncertainty about customer demand and the variability in production operations preclude successful in-line sequencing.

Level, sequential flow, on the other hand, is *adaptive and responsive to the customer*. It also ensures the even distribution of work for the factory and its suppliers. The rate of production and the product mix is much more even than with any other scheduling algorithm. This results in steady utilization of production resources and helps eliminate the "bullwhip" effect (Lee, Padmanabhan, and Whang 1997b).

The bullwhip effect refers to disruptive variability brought about by uneven customer demand. Consider a factory where customer demand suddenly becomes erratic. The average demand may stay the same, but the orders themselves are larger and sporadic. Perhaps the customer changes his purchasing policy and requests six months' worth of product at once rather than accepting smaller, weekly deliveries. To fill the larger order, the factory scrambles to produce at a higher rate. This places a sudden increase in demand on upstream processes. These upstream processes often overreact and overproduce to satisfy the increased demand, and they may build up extra inventory in preparation for a future increase in demand. The purchasing department notices a sudden increase in material usage and probably increases order sizes to support a presumed higher production rate. The suppliers may react to the larger orders by adding a shift, buying more equipment, or building a new plant.

To summarize this situation, each successive upstream process overreacts to accommodate the increasing unevenness in customer demand. The variability propagates and is amplified along the way. The bullwhip effect affects the entire value stream and stretches available resources. The farther upstream we follow the effects, the more extreme the effects are on the factory and its workers. Interestingly, all this is simply due to a customer changing his purchasing habits—an everyday occurrence for a factory!

Why is this called the bullwhip effect? How does a bullwhip work? Any slight movement of the handle initiates a wave of motion that travels in increasing amplitude, causing the tip to actually break the sound barrier. Similarly, unevenness in customer demand causes ever-growing fluctuation in upstream processes. Each process must somehow create a buffer against that fluctuation. The buffer can be excess capacity, longer lead time, or greater inventory, but the overall result is higher production cost or poorer customer service. "The choice for companies is clear: either let the bullwhip effect paralyze you or find a way to conquer it" (Lee, Padmanabhan, and Whang 1997a: 101-2).

Figures 8.2a and 8.2b illustrate two approaches to production scheduling. One is traditional "batch and push" production and the other is level, smooth production. These two approaches differ in how they use production resources and how well they satisfy customer demand. Although the example is simple, the comparison sheds light on very critical scheduling issues facing today's factory.

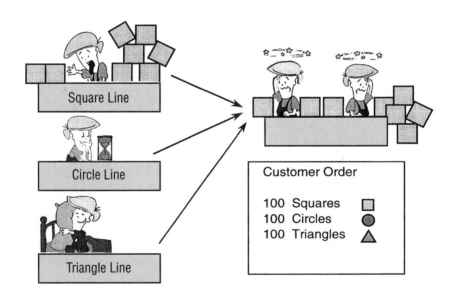

*Figure 8.2a* Traditional Batch and Push Production

Consider a factory that produces three items: squares, circles, and triangles. A customer buys 100 of each item every month. Unfortunately, the customer does not know *exactly* how many pieces of a given item he will order on any particular day. Therefore, the factory must schedule how many to produce and when to produce them based on incomplete information. Let's look at two alternatives to this production-scheduling dilemma.

The batch and push approach to production scheduling is illustrated in Figure 8.2a. Clearly the squares are being produced much faster than necessary to meet customer demand, and the resources for making squares are stretched to the limit. Meanwhile, the resources required to produce circles and triangles are poorly utilized. Squares that are not purchased accumulate in finished goods inventory. They are used to fill orders for squares during the production of circles and triangles. This approach puts excessive, sporadic demand on all of the resources!

Batch and push production may result in long production lead time and long customer waiting time depending on the batch sizes. The larger the batch or lot size, the longer the lead time. Assume there are no triangles in finished-goods inventory; if a customer requests one triangle while squares are in production, he must wait for the entire batch of squares and the entire batch of circles to be completed before he can have his one triangle.

Yet, many manufacturers still produce this way. Lot sizes are usually measured in the tens of thousands or hundreds of thousands, and we are familiar with some factories that produce in lots of multimillion pieces! Given such large lots, the production lead time may be weeks or even months.

There are rationalizations or "reasons" why a production manager might use batch and push production scheduling (Figure 8.2a). Perhaps the squares are easier to manufacture or have a higher profit margin. By making all the squares early in the month, the manager can "get ahead" on the monthly production quota. His department can be credited with the inventory, which is treated as revenue by most accounting systems. The department continues to appear profitable later in the month when the more difficult or less profitable jobs are produced.

Another possible reason is that changeover operations are lengthy, costly, or unpleasant. In this case producing in large lots may be viewed as more "efficient." Supervisors may favor larger lots to avoid lengthy changeover operations that hurt the production numbers for their shift. The point we are making is this: to be competitive, today's factory should look beyond individual processes or departments and avoid playing games with the balance sheet. We must look holistically at the factory's operations and pursue the optimization of the entire value stream.

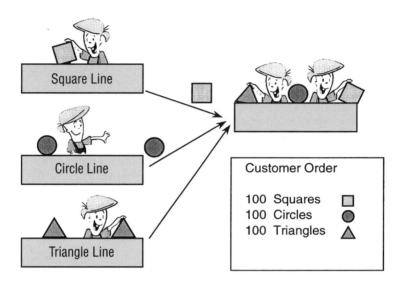

**Figure 8.2b** *Level, Smooth Production*

In Figure 8.2b, the production is smooth and even in terms of quantity and product mix. The tempo of production is matched to the rate of customer demand. Little or no finished goods inventory is required because the factory is making what the customers want in the proper proportion. The resources, such as equipment, raw materials, and workers, are never idle or overused. The production schedule is even, so the bullwhip effect does not transmit excessive variability to upstream operations or to suppliers. This production schedule (Figure 8.2b) also results in much shorter lead time. If the customer wants a triangle, the longest possible wait is the time required to produce one square, one circle, and one triangle. The waiting period or lead time is much shorter than in the batch and push case.

## Production Tempo

Today's factory sets the tempo or pace of production to the rate of customer demand. For example, if a customer uses a particular item at a rate of 2400 pieces per day, the factory should produce 2400 pieces per day. This principle, matching production to customer demand, can be used to establish production schedules, line rates, staffing levels, and delivery arrangements with suppliers. In practice this is done by using *takt* time. *Takt* is a German musical term meaning time,

$$Takt \text{ Time} = \frac{\text{Available Work Time}}{\text{Customer Demand}}$$

**Figure 8.3** *Production Tempo and Takt Time*

musical measure, or bar. Just as musical tempo is established by the conductor's baton, the tempo or pace of production is established by the rate of customer demand. The "beat" is ubiquitous throughout the value stream and determines the production rate at all stages of production (Figure 8.3).

To illustrate how this principle is applied, consider a one-shift operation, with 480 scheduled work minutes per day and a customer order for 1200 pieces per day. *Takt* time determines the production tempo. To compute *takt* time, simply divide the number of work seconds in the shift by the quantity demanded by the customer.

$$\frac{480 \text{ min.} \times 60 \text{ sec./min.}}{1200 \text{ pieces}} = \frac{28{,}800 \text{ seconds}}{1200 \text{ pieces}} = \frac{24 \text{ seconds}}{1 \text{ piece}}$$

According to the *takt* time, 1 piece should come off the production line every 24 seconds. The production rate is 1 piece every 24 seconds, and every operation up and down the value stream should produce at this pace.

Let's consider a more realistic example: a two-shift operation. Each shift is 8 hours long with a 30-minute lunch period and two 15-minute breaks. The customers are not satisfied with a single product and are demanding three different products. Since the customer is purchasing three different items, each item will have its own *takt* time. There is also an overall *takt* time for the entire plant.

8 hours – ½ hour (lunch) – ½ hour (2 breaks) = 7 hours per shift
2 shifts × 7 hours/shift × 3,600 seconds/hour = 50,400 seconds per day

There are 50,400 seconds available each day for production. The customer demand for the three items is shown below.

### Customer Demand:
■  840 Squares
●  420 Circles
▲  1260 Triangles

The *takt* time calculations are shown below:
■  50,400 seconds / 840 pieces = 60 seconds / square
●  50,400 seconds / 420 pieces = 120 seconds / circle
▲  50,400 seconds / 1260 pieces = 40 seconds / triangle

### Overall
50,400 seconds / 2520 pieces = 20 seconds / piece

On average, 1 piece is produced every 20 seconds. This is the overall *takt* time and is a result of the *total quantity ordered*. Given the *mix of orders*, this means 1 square is produced every 60 seconds, 1 circle every 120 seconds, and 1 triangle every 40 seconds. Figure 8.4 shows how the production line is sequenced in a factory that produces according to *takt* time.

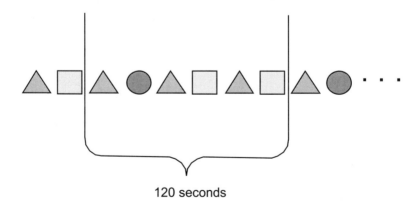

120 seconds

*Figure 8.4 Production Sequence Using Takt Time*

An observer standing at the end of the production line sees an item being completed every 20 seconds. A triangle is finished every 40 seconds, and, in between each triangle, a circle or square is produced. The interval between successive squares is not consistent. However, exactly 2 squares *are* produced every 120 seconds, which averages 60 seconds per square. The production tempo established by this sequence resonates throughout the value stream, requiring upstream operations to produce components and subassemblies at the same rate.

## Single-Piece Flow

Single-piece flow is one of the most important elements of just-in-time production. Single-piece flow is sometimes referred to as one-piece flow or continuous-flow manufacturing. We prefer the term "single-piece" because it clearly refers to discrete-parts manufacturing. Single-piece flow is the uninterrupted movement of material through the factory a single piece at a time. It is arguably the most efficient way to process material through a factory (Chapter 5).

Single-piece flow cannot be used in all manufacturing processes. For example, stamping, forming, molding, casting, plating, coating, and tumbling are often better suited for batch production. Also, equipment may not be appropriate for single-

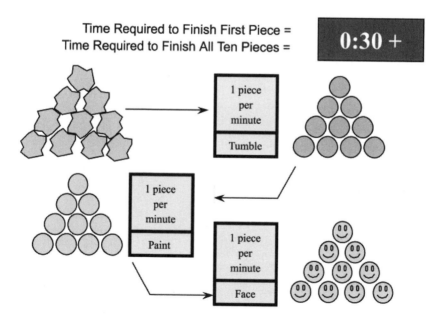

*Figure 8.5a* Batch and Push Production

piece flow if it produces multiple products, requires a lengthy setup, or operates at a speed much faster than *takt* time. Another situation that precludes single-piece flow from one process to the next is when those processes must be separated for safety or economic reasons. Where appropriate, however, single-piece flow results in unparalleled improvements in productivity, efficiency, flexibility, and profitability!

Single-piece flow is an extension of the sequential or product-oriented layout of Flanders and Wallering (Chapter 2). Their contributions to Model T production at Ford Motor Company include positioning equipment in the sequence of the production processes. In single-piece flow, the material is passed a single piece at a time from one process to the next.

Consider the batch production example shown in Figure 8.5a. In this example, it takes 1 minute per piece to tumble the rugged, raw material into smooth spheres. This process is performed in a batch, so it takes 10 minutes to process all 10 pieces. It also takes 1 minute to paint each sphere, so it takes a total of 10 minutes to paint all 10 spheres. Painting the faces also takes 1 minute per piece, so another 10 minutes are required to process them. All in all, it takes 30 minutes to complete the order plus any additional time needed to count, inspect, stage, and move the spheres from one process to the next. It is important to note that a mini-

**Time Required to Finish All Ten Pieces =** 0:12

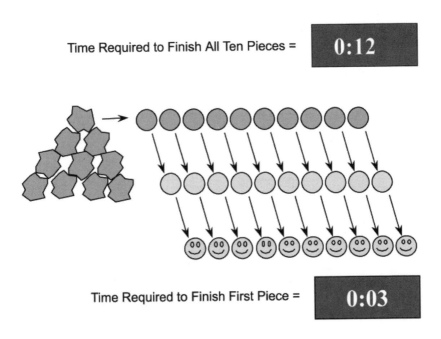

**Time Required to Finish First Piece =** 0:03

*Figure 8.5b Single-Piece Flow Production*

mum of 30 minutes elapses before even one piece is ready for the customer; not one piece can be sold until the entire batch is completed.

Figure 8.5b shows the same production operation using single-piece flow. Each sphere is moved immediately to the next station as soon as it is ready. The entire order is completed in 12 minutes, and it takes only 3 minutes for the first piece to be ready for the customer!

Response to customer demand is improved because the factory's production schedule is extremely flexible. Any changes can be made immediately because there is so little material that must work its way through the system. For example, if the customer decides to buy green frown faces instead of yellow happy faces, the first green face can be produced in 3 minutes!

Using single-piece flow, factory management must be willing to move material one piece at a time from one process to the next. This is impossible in a factory with a traditional floor layout based on functional process departments. Often processes such as welding or machining are in separate areas of the plant. They may be in different departments, each with its own managers, performance measures, and goals.

Sometimes "sequential" operations are in different buildings, different states, or even different countries! In these situations, management must be willing to take a broader view of the material flow rather than focusing on the efficiency of individual operations. Table 8.1 lists some common problems associated with batch and push production.

| "Batch and Push" Production Problems |
| --- |
| Excessive WIP |
| Extra conveyance |
| Multiple handling |
| Overproduction |
| Defective items |
| Difficulty solving problems |
| Lack of flexibility |
| Poor customer service |
| Long production lead time |
| Difficulty standardizing work procedures |

*Table 8.1* Batch and Push Production Problems

## Cellular Manufacturing

One way to achieve true single-piece flow for segments of a routing is to develop manufacturing cells. In these cells, the equipment needed to manufacture a product or product family is colocated and arranged in sequence. The workpieces are moved one by one from one production process to the next. The production rate in a manufacturing cell is set by customer demand, and *takt* time determines the pace. Obviously each operation in the work cell must have a processing time that is strictly less than the *takt* time. The layout of equipment and workstations is determined by the sequence of production. "Make one, move one" is followed rigorously in cellular manufacturing. If any one station produces at its own pace, accumulating the excess in a bin or even on a conveyor, the advantages of cellular manufacturing will quickly be lost.

Wherever possible, hand-to-hand transfers are recommended. There is no better way to ensure a proper balance. Another good way to transfer material is for one operator to load a machine and begin its cycle and for another operator to unload the machine. Conveyers should be avoided. They quickly become storage areas, and they hide imbalances that lead to an accumulation of material between operators. Queuing is forbidden! (Figure 8.6.)

*Figure 8.6 The Key to Cellular Manufacturing*

Since the pace of production is matched to the rate of customer demand, the use of slower, less expensive, or single-purpose equipment becomes an attractive alternative to expensive, computer-controlled, or multipurpose equipment. In traditional factories, managers have recognized for a long time the efficiency gains from having an operator run more than one machine. In today's factory, however, operators need to be familiar with many different processes so they can perform different tasks and operate more than one kind of equipment. Operators should be "multifunctional." Some useful guidelines to keep in mind for cell design are:

- no overproduction allowed!
- consider ergonomics and material presentation
- workers should stand and move among the stations
- use fixtures and clamps instead of adjustment
- do manual work in parallel with machine work
- never pass on a defect!

When designing a manufacturing cell, operator activity and material movement are much more important than the shape or appearance of the cell. A cell may be linear, L-shaped, U-shaped, semicircular, or virtually any other conceivable shape. Some shapes have inherent advantages over others, but the design of the work cell should never be based on a preconceived visual ideal.

Figure 8.7 is a layout of a typical U-shaped cell. Notice that the workers are inside the U. There are several reasons for this. First, having the workers on the inside reduces the distance traveled by production material on its way through the

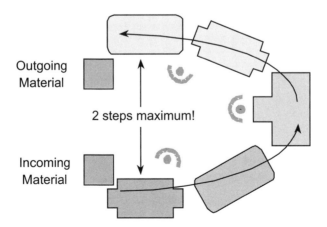

*Figure 8.7* A Typical U-Shaped Cell

151

cell. Similarly, the distance the operators travel is minimized. Also, proximity promotes teamwork; when workers are near each other, it is easier to see the next worker as a "customer."

We have been in small factories where the operators of two successive operations didn't know each other's names or where their workstations were located. In such cases, we recommend making introductions to reinforce that there is an actual person—a human being—associated with every job in the factory. Meeting each other is a good way for operators to begin developing a customer-supplier attitude.

Having workers inside the work cell also makes multifunctional work possible. Operators can step back and forth across the cell to perform multiple tasks. Another advantage of the U-shape is that production output can be adjusted easily by changing the staffing level. When customer demand slows, staffing can be reduced. When demand increases, more operators can be assigned to the cell.

Cellular manufacturing is a popular and highly visible lean manufacturing technique. It is often adopted by manufacturers who misunderstand its intent. Consequently, the benefits of cellular manufacturing are unrealized. We offer two cautions with the hope that they may avert disappointment. First, take care not to process material through the cell in batches. Batch production seems natural for many operators, but it results in true worst case performance (Chapter 5). Second, never allow the output from a processing station to accumulate within the cell. The material flow and the processing time should be governed by the *takt* time.

## Traditional Approaches to Line Balancing

Traditional industrial engineering techniques are excellent for determining the number of workers required to perform a given set of production tasks. (These well-known techniques can be found in any basic industrial engineering textbook.) Often, however, there is inequity in the effort or time required for individual operators to do their jobs. The challenge of equalizing each operator's workload is called line balancing. Figure 8.8a illustrates the time required for four operators to complete their jobs. Notice that the time required for Operator #2 to complete his tasks is longer than the *takt* time. Since the *takt* time is the pace of customer demand, this "cell" cannot satisfy the customer.

This line can be "balanced" in the traditional sense by transferring work elements until all operators have approximately the same workload (Figure 8.8b). With these adjustments, the work can be completed within the *takt* time, and the customer demand can be met. Unfortunately, the pace of production is much faster than the tempo established by *takt* time. This leads to overproduction and all the associated wastes (Chapter 7).

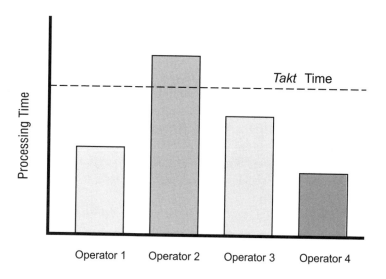

**Figure 8.8a** *Operator Imbalance*

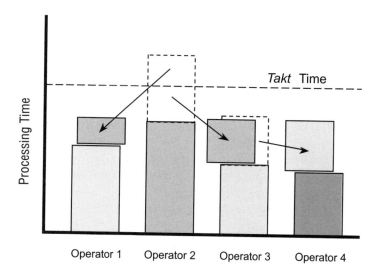

**Figure 8.8b** *Traditional Line Balancing*

There is another approach to balancing a production line; it attempts to incorporate lean practices into the traditional balancing method discussed above. To illustrate this approach, let's reconsider the example presented above (Figure 8.8a). This time operation #2 is targeted for an improvement effort or even a *kaizen* event. The goal of the "improvement effort" may be to reduce the time required to complete operation #2. Then the work elements are redistributed, and the operations are balanced in the traditional manner. The end result is that less total time is required to complete the combined operations. Lean improvements (such as *kaizen* and continual improvement) can be claimed. Standard labor reports indicate that the production "efficiency" of this isolated area is higher. As an efficiency exercise, the balancing project seems to be successful and commendable. What is wrong with this picture?

What is the true lean approach to balancing a production line or a work cell? It is to match the pace of production to the rate of customer demand. This central objective of lean philosophy is strongly supported by factory physics. What has happened in this case? The production rate has outstripped customer demand by an even larger margin than in the traditional line-balancing example (Figure 8.8b). Overproduction is a certain result, bringing with it an entire battery of problems such as high inventory, long cycle time (total time material spends in the factory), high production cost, and so on. Total system efficiency is not achieved by maximizing the efficiency of individual pieces of the system.

## A Lean Approach to Line Balancing

Lean manufacturing attempts to utilize human effort at a very high level but never at 100%. In fact, today's factory does not aim for 100% utilization of any resource, whether it is equipment or employees. In the lean approach to line balancing, work elements are distributed so that the amount of time required for each worker to complete his tasks is slightly less than the *takt* time. This may result in a worker being underutilized (Figure 8.8c). In today's factory this is seen as an opportunity and an advantage, not a shortcoming.

This approach to line balancing has three advantages. First, if the cell is out of balance, it is obvious that one worker is underutilized and that the production line needs further improvement. Second, the underutilized worker has time to do work that only humans can do, such as devise better work methods or process improvements. Later, when improvements are implemented, it may be possible to remove a worker from the production line. This reduces the cost of producing the item, and the worker becomes available to do other things. The worker may move to another work cell, train for another job, acquire new skills, become a shop floor team leader, or join a "team" that has responsibility for improving production processes throughout the factory.

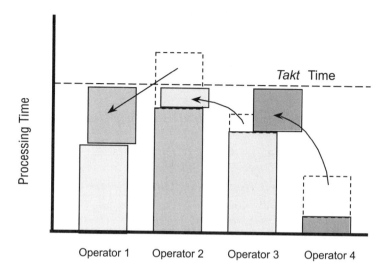

**Figure 8.8c** *A Lean Approach to Line Balancing*

A third advantage is pacing production to meet customer demand; there is no overproduction, and there are none of the associated wastes. The apparent inefficiency of having an underutilized worker is completely offset by the efficiency with which production material flows through the plant. This streamlined material flow brings many advantages such as improved response to the customer, higher quality, shorter production lead time, shorter cycle time, less inventory, and higher profitability. In today's factory, the localized efficiency gains from traditional line balancing are eclipsed by the factory-wide benefits from streamlining material flow.

## Just-In-Time Delivery

A corollary to just-in-time production is just-in-time delivery. To clarify how the two practices relate, we will discuss two approaches to receiving material from suppliers.

In the first approach, the factory accepts delivery of production material in large quantities on an infrequent basis. By "large" we mean that enough material is delivered to satisfy production requirements for several weeks or even months. This could be in response to the "economic" order quantity from the company's MRP system, or it could be because of real or perceived shipping costs. Whatever the reason, most manufacturing companies still receive sporadic and infrequent delivery of items from distant suppliers.

Material handling departments prefer larger lots, larger containers, and fewer deliveries. This means fewer trucks or rail cars to unload, and fewer trips to deliver material to the production line. For example, material handling managers who have an incentive to reduce head count and increase "efficiency" may lobby strongly for larger container sizes. This improves the operation of the material handling department but does it at the expense of the overall production system. Quality assurance departments may also have an incentive to inspect material in larger lots since there is an associated economy of scale.

These functional departments with their own specific incentives are making good decisions based on their performance measures and targets. Unfortunately they are creating severe problems for the factory as a whole. Large lot sizes and infrequent deliveries contribute enormous variability to the production flow (Figure 8.9a). This causes longer customer lead times, reduced flexibility and responsiveness, and decreased profit.

Fortunately for today's factory the concept of frequent deliveries is beginning to take hold. Factories are accepting regular and frequent deliveries of production material in much smaller quantities from suppliers. Additional shipping costs incurred are offset by the lower financial investment in inventory. We find that purchasing managers are surprised by the willingness of suppliers to deliver in this

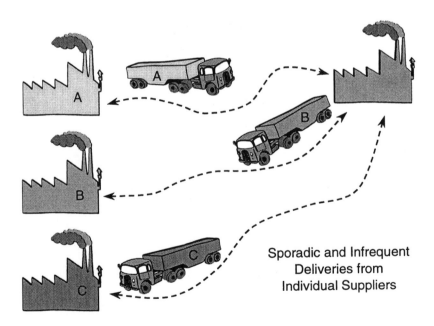

Sporadic and Infrequent
Deliveries from
Individual Suppliers

*Figure 8.9a* *Traditional Conveyance*

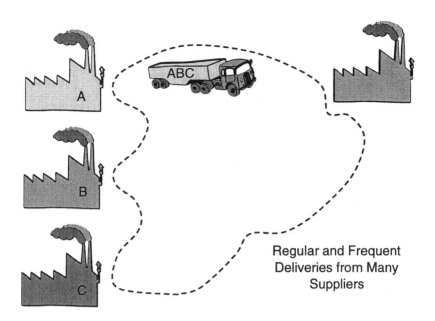

Regular and Frequent
Deliveries from Many
Suppliers

*Figure 8.9b* Mixed-Load Delivery

way. Surely it is much easier for suppliers to plan their production when they are delivering 6,000 pieces every week rather than 60,000 pieces at uncertain and infrequent intervals, often without sufficient notice. The minor inefficiencies in functional departments such as purchasing, material handling, and quality are outweighed by the benefits of frequent deliveries. We have noticed that once the people in these functional groups understand the benefits, they find ways to improve their own processes so that the increased delivery frequency poses no problem at all!

Another approach to material conveyance that is gaining acceptance is mixed load pickup and delivery (Figure 8.9b). In this scenario an individual truck makes stops at many suppliers, picks up the items, and delivers them to the factory. This is sometimes called a "milk run" because of its similarities to the milk routes common in the 1950s and 1960s, when the milk truck delivered to individual homes and replaced whatever items had been consumed such as milk, cream, eggs, and butter. An entire industry has grown up around the need to supply the factory with production material, and many companies are using third-party logistics providers to perform this service.

It is important to distinguish JIT production from JIT delivery. JIT delivery is a necessary but insufficient condition for JIT production. It is possible to deliver on a just-in-time basis without changing the way production in the factory is managed. Without adopting JIT production along with JIT delivery, the "time" in just-in-time delivery would be a mystery: just in time for what?

We are familiar with an automotive assembly plant that adopted JIT deliveries using "milk runs." Every day, one day's worth of supplies arrived at the plant "just in time" to satisfy the daily requirement. One Monday morning, the plant painted a whole week's worth of boysenberry colored cars. The problem was that the door handle supplier shipped one day's worth every day. So a week's worth of boysenberry cars hit the line with only one day's worth of boysenberry door handles. Additional door handles had to be flown by airfreight from the supplier so the cars could be completed and ready for delivery. JIT delivery without JIT production most certainly increases cycle time and inventory. Quality, customer service, and profitability suffer, and the benefits of just-in-time production cannot be realized.

## Summary

The just-in-time production philosophy is an excellent way to make what the customer wants when the customer wants it. There are several practices that support just-in-time production. These practices reduce the *amount of time and the variability in the time* required to process material through the value stream. The concept of level, mixed production helps the factory make what is needed when it is needed. This reduces lead time and the time material waits to be used or sold. Customer demand sets the pace of production. Single-piece flow, small-lot production, and cellular manufacturing reduce the amount of time material spends in the value stream. In fact, where applicable, single-piece flow is the fastest way possible to process material through a system. Mixed conveyance reduces variability by delivering small lots from many suppliers frequently rather than delivering large lots from a single supplier infrequently. This contributes to an overall reduction in the time material spends in the factory because smaller lots of material are consumed much more quickly than larger lots.

The secret of why just-in-time works in today's factory lies in its ability to shorten cycle times and reduce variability. To do this effectively, suppliers and upstream processes must somehow know what the factory needs and when it is needed. Consequently, the timely and accurate flow of information from today's factory to its upstream operations and suppliers is of paramount importance in the implementation and execution of JIT.

*Chapter 9*

# Pull Production Control

## Introduction

A pull system is a simple and effective way for today's factory to manage its material and control its production. A pull system responds in real time to the status of the factory; it is self-correcting and self-regulating. A pull system is essential to successful JIT production, but it is by no means a zero inventory policy. The premise of pull is to replenish material that has been consumed, so some material is always present.

A pull system can be implemented early in a factory transformation, and the benefits are realized instantly. The most notable advantages of a pull system are that stockouts are eliminated and inventory drops markedly. Other benefits include a streamlined flow of material and information, shorter cycle time and lead time, greater flexibility, higher revenues, lower production cost, and ultimately higher profit.

In this chapter we explain the concept of pull production control using a nonmanufacturing analogy, and we define it in clear, concise terms. We also offer compelling evidence of its ability to increase profit, and we discuss implementation issues. We also answer several common questions about pull systems in general.

## Rivers and Dams: A Pull Production Analogy

Production control concepts can be illustrated clearly using a river and dam analogy (Figures 9.1a and 9.1b). In this analogy, water flowing down a river represents production material flowing through the value stream. Dams along the river control the flow of water. They represent stages in the production process. Controlling the flow of water in the river is analogous to controlling the flow of material through a production routing. The objective is to release enough water at each dam to supply the next downstream community.

The community at each level along the river needs water for irrigation, recreation, transportation, and electrical power generation. If the water level is too low, the community suffers and cannot carry on as usual. If the water level is too high, the river floods and causes hardship and disruption. Each community depends on the upstream dam to supply water as needed.

In our river analogy we use two approaches to control water flow: a traditional *push* system and a *pull* system. When a push system is being used, a central control office located 500 miles away sends a release schedule once a month to the dam operators along the river (Figure 9.1a). The control office relies on a sophisticated computer algorithm to calculate and "control" the release of water based on past weather bureau records and recent forecasts. When the schedule indicates that water should be released, the dam operator opens the gate, and water slowly flows to the next community. The computer has been programmed to provide safe and sufficient flow throughout the river system, so the release of water occurs without regard for the readiness of the downstream community.

This system would work in an ideal world. In reality there are complications that arise because of variability and unsatisfied assumptions made by the controlling algorithm. Unforeseen difficulties, such as absenteeism, malfunctioning spill-

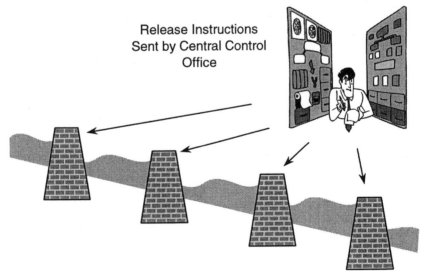

Release Instructions
Sent by Central Control
Office

*Figure 9.1a* Centralized Flow Control

ways, unexpected rain or drought, or a sudden demand for more water or electricity could render the system ineffective, unsatisfactory, and even dangerous. Unexpected occurrences contribute uncertainty and variability to the system.

This approach is analogous to a push production control system that schedules the factory based on orders or forecasts using material requirements planning (MRP) or a related algorithm. In a push system, the orders or forecasts are analyzed, and production material is procured and scheduled to arrive when needed. The factory is scheduled to produce necessary components and subassemblies for timely production of the finished goods.

This type of system would work *extremely* well if demand were perfectly predictable and within the capacity constraints of the factory. In such an environment, an MRP-type system would actually result in just-in-time production (Hopp and Spearman 1996: 157). As our river and dam analogy indicates, the problem comes from unexpected problems and unsatisfied assumptions that are unavoidable in any manufacturing environment. We believe most manufacturing managers will agree: *the forecast is always wrong, schedules always change, and nothing goes according to schedule anyway.*

In a pull system, the release instructions do not come from a centralized control office. Instead, the water levels are monitored constantly (Figure 9.1b). When the water level at Dam 1 falls below a predetermined threshold, the release signal or *request for replenishment* is sent to the next upstream dam (Dam 2). As water

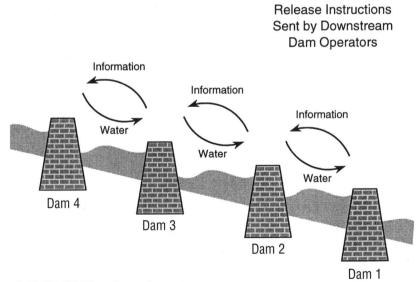

*Figure 9.1b* "Pull" Flow Control

flows from Dam 2 toward the next downstream community (Dam 1), it is possible that the water level at Dam 2 will drop below *its* threshold. If so, a signal is sent to Dam 3, the next upstream dam, to release water from its reservoir. In this way, information flows upstream from dam to dam, and water flows downstream in just the right amount to replenish what has been consumed.

This system works successfully regardless of unforeseen difficulties. The water does not fall below the specified minimum level, and the system cannot flood with excess water. What happens if an unexpected problem prevents an operator from releasing water and the level does not drop as expected? The replenishment signal is not sent to the next upstream operator, and, consequently, no additional water is received. There is no danger of flooding while the problem is being fixed. What happens if there is an unexpected surge in demand for water by downstream communities? Signals are sent upstream more frequently, and the water is replenished more often. A pull system like this is self-correcting and self-adjusting in the face of unforeseen problems.

A river that is controlled by monitoring water levels is analogous to a production system that is controlled by pull signals. The objective is to provide all processes with real-time information about the timing and quantities of material required (Yanagawa, Miyazaki, and Ohta 1994: 163). A pull signal is simply a request for replenishment. It is a message to an upstream operation that says, "I have depleted my production material, and I need more." A pull signal is not based on a schedule or a forecast; it is a reaction to material that has been consumed. Production material is released only to replenish what has already been used. In contrast, a push system releases material to the next process to satisfy a pending demand.

*Pull is responsive; push is anticipatory.* A pull system directly controls WIP and monitors throughput. A push system monitors WIP and attempts to control throughput (master production schedule) (Spearman and Zazanis 1992: 522).

## The Problems with Push

For the past 30 years, manufacturers have been able to use computers to plan their production material requirements and generate their master production schedules. The algorithms used in these programs were, and still are, based on the "push" approach to production control. The systems became known as Material Requirements Planning (MRP) systems for obvious reasons.

In 1972 the American Production and Inventory Control Society (APICS) began its MRP crusade. By 1981 as many as 8,000 MRP production control systems had been installed in the United States. In a 1982 APICS-funded survey, 60% of MRP users reported that they either had a modestly effective but partially implemented system or that they had a marginal system providing little benefit to the company (Anderson, Schroeder, Tupy, and White 1982). Other researchers using other data point out that in 1984 "ninety percent of MRP users are unhappy" (Whiteside and Arbose 1984). As time went by, companies were still not realizing the promised benefits. "Although MRP is a well-developed data management system, it has not lived up to expectations for increased system performance" (Deleersnyder, Hodgson, Muller, and O'Grady 1989: 1079). Five years later, more than 70,000 systems had been implemented, and two-thirds of MRP users were dissatisfied with the product (Blood 1994).

Why isn't the manufacturing community realizing the promised benefits of MRP and related systems? The discrepancy between expected and actual results can be explained quite easily. MRP is an excellent mechanism for determining what materials and components will be required and when they will be needed to achieve a specific, user-defined production schedule. In fact, it is almost tautological that material requirements planning (MRP) is ideal for planning material requirements (PMR)!

The hitch comes when MRP is used to control production rather than to plan materials. For the purpose of production control, the assumptions and models underlying MRP are fundamentally flawed. For example, MRP algorithms assume incorrectly that the time required to procure an item from an internal or external supplier is independent of the quantity ordered. MRP also assumes that the time required to produce an item is independent of the status or loading of the plant when, in fact, the opposite is true! It is well known to factory managers (and can be proven with factory physics) that *the time required for production material to be processed through a factory does depend on the loading of the plant.* (The exception is a plant with infinite capacity.)

MRP makes erroneous assumptions about how a real factory actually runs. The result is parts shortages and late deliveries to customers. Many materials managers compensate by inflating the lead times in the system in an attempt to have adequate material and components available for production. Unfortunately, this compensatory "solution" has a snowball effect. The increased lead time allows more work to be issued to the factory; consequently, more jobs are in process at any given time. This, in turn, increases congestion and cycle time causing even more dramatic increases in lead time!

So the attempt to use MRP for efficient, just-in-time production control causes the factory to become mired in its own inventory. Traditional MRP and batch manufacturing are very inflexible. What does this mean for today's MRP users? "The net effect is that MRP, touted as a tool to reduce inventories and improve customer service, can actually make them worse" (Hopp and Spearman 1996: 175).

Recently a growing number of enterprise resource planning (ERP) systems have become available. "While ERP systems are useful for many manufacturing functions, MRP logic remains imbedded in the planning process" (Layden 1998: 51). The latest computer software designed to improve operations and efficiency is advanced planning and scheduling (APS). John Layden, a founding father of APS, describes these systems as "realistic" and "based on real-world factors." However, they have a remarkable inability to handle very ordinary, "real-world" complications with which today's factory must contend:

- unstable production schedules
- production modes other than batch and queue
- production requiring optimal sequencing
- changes in factory status
- changes in customer demand.

In factory demonstrations we often use a hands-on assembly simulation to contrast push and pull production systems to managers and shop floor workers. After we run the simulation using a reasonable push model, we ask the audience to offer recommendations. A frequent suggestion is that a more sophisticated computer model would improve the push system by generating a better production schedule; this is the exact solution many companies are trying today. However, the latest software is definitely not a solution. "New software can make things worse rather than better" (Nahmias 1997: 380). This is particularly true if the new software is based on the premise of push!

## The Power of Pull

*Pull systems are essential to just-in-time production*, but they are not zero inventory systems. Some inventory must be in the system at all times. The amount of inventory is *controlled* by the system; it fluctuates between preset minimum and maximum levels. The levels are established based on usage, cost, lead time, frequency of replenishment, and the desired customer service level. Setting the inventory levels also requires judgement so that any other relevant factors are taken into consideration.

Pull systems have a *cap* on the amount of work-in-process (WIP) that can exist in the system. This *WIP cap* reduces congestion and improves production flexibility. Another benefit of a WIP cap is shorter cycle time; this is because cycle time is directly proportional to WIP for a given throughput rate. Shorter cycle time has several important advantages (Spearman and Zazanis 1992: 524):

- shorter lead times
- lower reorder points
- greater production flexibility
- reduced risk
- lower investment.

## Variability

In a push system there is no direct control of WIP, so inventory levels vary. In fact, inventory levels in a push system are *highly* variable. One workstation may be flooded with WIP while another is low or lacking in material. The variability of WIP affects the variability of cycle time since WIP and cycle time are related. The high WIP fluctuations in a push system cause high variability in cycle time. *A longer and more variable cycle time means that longer lead times will be needed* to achieve a given customer service level (or percentage of on-time deliveries).

On the other hand, the low variability in cycle time afforded by a WIP cap means that much less inventory accumulates between stages of production, and less "idle" inventory is needed in finished goods. A managed amount of inventory is placed deliberately at specific locations along the value stream. These locations are typically where production rates are mismatched, where equipment requires a long setup, or where equipment is subject to frequent breakdowns, in other words, where processing time variability is high.

Looking at the entire production process, much less inventory is needed to support a pull system, and this means that less capital must be invested. One department in which we implemented a pull system lowered WIP by one-third in four months; the new production control system freed over $100,000 in capital.

Some operations researchers state that the power of a pull system lies in its ability to cap the WIP, in other words, to restrict in-process inventory to a predefined range. Their assertion may be true from a strictly mathematical point of view; a WIP cap certainly provides many quantifiable advantages.

Other advantages are less easily quantified. For example, customer orders are rarely stable, and they often fluctuate widely. One manager at an automotive stamping plant admitted that 80% of all customer "releases" changed between any given Friday and the following Monday. A pull system tends to *moderate these demand fluctuations* that "perturb" (mathematically speaking) all factories (Kimura and Terada 1981).

We maintain that a pull system provides advantages for line management. Pull systems are *highly visible*. They can be implemented using color-coded bins, cards, or tape. These visual organizers also *physically organize* inventory by the size of the bin, bag, box, tote, rack, tub, or designated floor space. The visibility and organization of a pull system make a production line easier to manage. There is clear control of the amount of inventory at each location, and shop floor workers can operate pull systems with little or no management intervention.

Another advantage for line management is that pull systems are *reactive*; they respond immediately and dynamically to unexpected variability such as quality problems, equipment failure, or sudden changes in line rate (Muckstadt and Tayur

1995: 141). This self-correcting feature of a pull system makes it inherently easier to control than a push system (Spearman and Zazanis 1992: 530).

We have noticed another practical advantage of pull systems. Pull systems reinforce the idea that *the next process is a "customer."* Supplying a "customer," an actual person rather than a pile of WIP, provides strong psychological incentive to see that parts are correct and delivered as soon as possible. Too often an operator produces items according to a predetermined schedule and has no way of knowing when, or even if, the parts are used. This does not promote the fact that the next process is a customer. On the other hand, pull systems reinforce that people are providing material and components to each other throughout the factory. We recommend that the people involved in a pull system get to know each other. This personalizes the customer–supplier relationship within the factory.

The power of pull is its ability to keep the factory supplied with production material in the most reliable and efficient way possible.

## The Robustness of Pull Production Control

We have established that a pull system results in less congestion, greater flexibility, lower production cost, shorter cycle time, and shorter customer lead time. These are compelling reasons for using pull production control. However, before any production control strategy is adopted, it should be proven robust and insensitive to the uncertainty and variability inherent in any factory. In today's factory managers need to be confident that their production control system will perform properly in the presence of variability and uncertainty.

How can we compare the robustness of push and pull production control systems? One way is to develop a model of a simple production process and analyze its cost performance under both production control policies. Consider the Pushmepullyou Manufacturing Company. Pushmepullyou has a production line with four identical stations and an average processing rate of 1 piece per hour. Pushmepullyou receives a profit of $100 per piece produced but incurs a holding cost of $1 per hour per piece for every piece of in-process inventory. Pushmepullyou developed a cost model to approximate the profit generated by the production line. This model (shown below) is admittedly simplistic, but it helped the Pushmepullyou managers learn what they wanted to know and it illustrates our point very clearly:

Profit per hour = throughput × profit per piece − holding cost × WIP

Pushmepullyou computed the maximum profit for a push system and a pull system. For this comparison they chose a simple kind of pull system known as a constant work-in-process (CONWIP) system. As might be expected, the CONWIP system maintains a constant amount of WIP in the system at all times. To make this comparison between the push and CONWIP systems Pushmepullyou made some simplifying assumptions, namely that processing times are exponentially distributed and that transportation time is negligible. Since the profit equation has two unknown variables, Throughput and WIP, Pushmepullyou had to understand how these two production variables relate to one another. CONWIP systems control WIP. Therefore, throughput is a function of how much WIP is in the system. For this case, the relationship is as follows:

$$\text{Throughput} = \frac{\text{WIP}}{\text{WIP} + 3}$$

Push systems control throughput by controlling the amount of work that is released. A well-known result from queuing theory (shown below) can be used to compute the amount of WIP in such systems.

$$\text{WIP} = \frac{4 \times \text{Throughput}}{1 - \text{Throughput}}$$

(Notice that WIP becomes infinite if the factory tries to release work at the average processing rate (1 piece per hour). This means that the release rate must be strictly less than the average processing rate. This is not a special case but a requirement of all production systems.)

For their production line, the maximum profit under a CONWIP system is $68.35 per hour. The maximum profit under the push system is only $64.00 per hour. Pushmepullyou concluded that a pull system is more profitable than a push system when each is operating at its optimal point. Before we discuss this "optimal point," recall that a pull system controls WIP and a push system controls the rate at which orders are released to the shop floor.

The optimal point is where profit is highest. In a push system this means that the rate at which orders are released is high enough to ensure excellent throughput but not high enough that too much WIP accumulates. If orders are released too slowly to the shop floor, the throughput rate will be too low to generate much profit. If shop orders are released too rapidly, the production line will become mired in WIP. In a pull system the optimal point is where the WIP is high enough to sustain an excellent throughput rate, but it does not accumulate and become a significant financial investment. High WIP means high holding cost (inventory, storing, moving, staging, re-packing, etc.). If WIP is set too low, the throughput will also be low. This means less revenue for the factory and less profit. So, Pushmepullyou decided that a pull system would be more profitable than a push system, even when each is operating at its highest profitability.

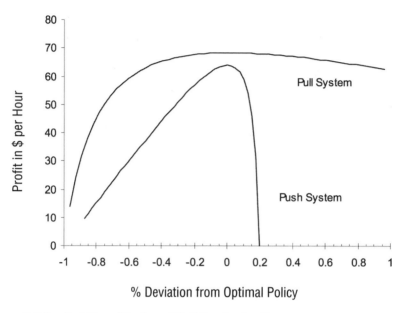

**Figure 9.2** *Profitability of Push and Pull Production Systems*

This is interesting, but we still need to address the issue of robustness. For example, how does a push system behave when the orders are released to the shop floor too slowly or too rapidly? How does a pull system behave when the WIP levels are set too high or too low? To answer these questions we must compute the profits generated by the push and pull control policies when they are operating above and below their optimal points (Figure 9.2).

Figure 9.2 shows clearly that a pull system is more profitable than a push system when each system is operating at its respective optimal point. What is even more interesting is that *pull production control outperforms the "optimal" push system even when WIP levels are set 40% too low or 80% too high!* Conversely, the profitability of a push system declines steadily if the order release rate is too low and drops sharply if the release rate is too high. In fact, the "profit" becomes negative if the release rate is more than 20% above optimal!

This is very disconcerting! Many factory managers experience pressure to maximize throughput and increase utilization of labor and equipment. Such pressure can easily lead to a decision to release too much work rather than risk having an underutilized operation.

The mathematical proof shows that pull systems are less sensitive than push systems even when the WIP levels chosen are not the best. This means Pushmepullyou can make gross errors in calculation (or judgement) and still have a production control system that works better than a push system would work at its very best. In light of all this evidence, Pushmepullyou decided to implement a pull production control system.

However, one question remains. How much inventory should be kept to ensure that the pull system operates efficiently? In practice we usually recommend setting the *initial WIP levels* slightly too high. While this approach may cause some lean purists to wince, it *is* an excellent way for companies to introduce a pull production control system. Setting the levels slightly too high is risk free. This is important because the pilot pull project must succeed to gain the confidence of everyone in the factory. Setting the WIP levels slightly high also makes it easier for people to accept the pilot pull system. It also creates less anxiety for the workers and managers. Implementation, acceptance, and use of pull production control are initially far more important than trying to completely maximize the efficiency of the new system. At a later time the WIP levels can be reduced easily.

There are several analytical techniques for computing the proper inventory levels for a pull system, and we discuss some of these later in this chapter. However, in the early stages of implementing pull production control, do not be too concerned about determining precise pull quantities. A seat-of-the-pants pull system will almost certainly be more profitable than a push system.

We do highly recommend that workers and managers be included when WIP levels are set. They are aware of production and scheduling complications that need to be considered. Also, their uneasiness will subside if they approve the WIP levels that affect their job responsibilities. We often encourage people with friendly reminders such as, "pull systems are extremely forgiving" and "you can't break a pull system!"

## Operators and Pull

Most operators want to be productive and do what they can to strengthen their company's competitive position. Being productive requires having *usable and acceptable* production material available where it is needed when it is needed. So, obviously, the quality of production material is a concern to operators when a pull system is implemented. When a pull signal is transmitted upstream, material has necessarily been consumed downstream, and the downstream operation needs more material. If defective material is sent to the next process, production may be interrupted or even stopped.

Recently we were implementing a pull system in an aerospace electronics factory. Operators in the assembly department were accustomed to receiving components in quantities slightly higher than their shop orders. This allowed the entire order to be completed even when a small percentage of the components was defective. If a large percentage were defective, operators could not complete the shop order until replacement material was received. The policy of our new pull system was that when the department's supply bins were emptied, the storeroom or fabrication department would refill them. One worried assembler asked us how she would get replacement material if as much as 50% of the material was unusable. What would she do the second half of her shift? We explained that if she had to scrap 50% of the production material, she would be "consuming" it at twice the expected rate. The pull signal (the empty bin, in this case) would be returned to the previous operation twice as fast, and her material would be replenished twice as often; she would always have enough material to complete a shop order! Her concern gave us an opportunity to discuss the self-correcting nature of a pull system. This helped her gain confidence in the pull system. She kept a log for a few weeks until she realized that she would always have components with which to work.

We find that most production workers and managers like using a pull system. With a pull system the operators know the status of the factory at all times and how their particular parts are being used. They are confident that they will have the material they need when they need it. Production is not interrupted because of parts shortages. Consequently, they are not pressured to make up for lost time by working furiously. On the other hand, the operators are not under pressure to produce items that aren't needed; when the downstream operation needs items, the signal is sent.

## Implementation

Pull systems manage the flow of information *and* material. Therefore, we need *a method for conveying information* and *a place from which to pull the material*. There are many ways to send a pull signal. One common approach is to use the containers themselves to convey both the material and information. If containers are used as the signal, the containers should be labeled with information such as part number, quantity, and location. This makes it easier for material handlers to know what to do with the container. The arrival of an empty container at an upstream processing station is a clear and unambiguous signal that cannot be ignored: the parts that were in this container have been used, and we need more! This approach is very intuitive and visual, and operators seem willing to support the use of containers as signals.

Another common approach is to use cards, often called *kanban* cards or simply *kanban*. (*Kanban* is a Japanese word meaning signboard.) Cards can be stacked, sorted, and handled easily. They can also be read by bar code scanners, faxed to suppliers, or attached to generic bins. Unfortunately, cards are also easily lost, misplaced, forgotten, or thrown away. We recently observed the failure of a pull system at an automotive assembly plant near Detroit. The failure was due largely to lost cards.

Pull signal systems can be quite creative. We are aware of systems that send pull signals via golf balls, ping-pong balls, clothespins, clips, designated floor space, dedicated carts, electronic signals, and the Internet! Whatever method is chosen, *the pull signal must initiate some action*!

There are two types of action that can result from a pull signal: material must be moved or conveyed, or something must be produced. If the appropriate action does not take place, the pull system will fail. Pull signals are not subject to interpretation: *action should take place*. Material has been consumed at a downstream process, and that material needs to be replenished.

The place from which we pull the material is a location we call a *marketplace*. This term brings to mind a neighborhood grocery, a convenient place where we can get the supplies we need. The grocery has a small quantity of many different items. When items are purchased, the shelf is replenished with more of that particular item. So, like the corner market, a production material marketplace is stocked with a small amount of raw material and various components needed for production.

The marketplace holds raw material, purchased components, or items and sub-assemblies that are produced in the factory. When a signal is received from a down-stream operation, the appropriate material is taken from the marketplace and sent to the workstation requesting the material. This partially depletes the marketplace. This depletion is communicated to the workers responsible for keeping the marketplace replenished with that item. If it is a purchased item, the signal goes either to purchasing or directly to the supplier. If the item is produced in-house, a signal is sent to the appropriate processing station, and that station produces the quantity necessary for replenishment. In this way, the marketplace is self-managed and always has a sufficient, but not excessive, quantity to meet production needs. It is also possible for a signal to be sent directly from the station that uses the material to the station that produces it, bypassing the marketplace altogether. However, this would require quick response because material would have to be produced immediately to replenish the supply.

Once it has been decided how the pull signals will be transmitted and where the marketplaces should be located, two more decisions must be made: when material should be conveyed, and how much material should be conveyed.

When should material be conveyed? Material should be replenished whenever the amount at a workstation becomes so low that the ability to meet production requirements is threatened. Remember that a pull system is not a zero inventory system. We do not wait until the inventory is depleted before replenishing it. Setting the signal threshold too low may cause production to shut down while workers wait for the needed parts; this results in lower throughput, longer cycle time, longer customer lead time, and lower profitability.

The replenishment frequency depends on the amount of material contained in each delivery. In general, frequent deliveries of small quantities are preferable to infrequent deliveries of large quantities (Chapter 5). Frequent deliveries may be less efficient for individuals or functional departments, such as receiving or material handling, but the efficiency of the overall factory improves. Frequent deliveries of smaller lots are far more efficient from a holistic perspective. (Remember that the variability is roughly proportional to the size of the transfer lots. This means frequent deliveries of smaller lots contribute far less variability to the production system.)

How much material should be conveyed? In the first English language article ever written about the Toyota Production System, Toyota suggests that the container capacity should not be more than 10% of the daily requirement (Sugimore, Kusunoki, Cho, and Uchikawa 1977: 561). For factories where pull production control is just getting started, we suggest that the container quantities be no greater than 25 to 50% of the daily production requirement. This may not be as "lean" as it could be, but it reinforces the new production control methodology without overtaxing any resources.

In the long run, it is the confidence and participation of people, not bin sizes, that determine the success of the pull system. Modifying container quantities is simple, and it can be done later in the implementation process. We prefer to focus first on how the workers perceive the production process, how their material gets replenished, and how production is controlled.

## Pull Quantities

Sugimore, Kusunoki, Cho, and Uchikawa (1977) present an equation for computing the number of *kanban* cards for each item in a pull system.

$$y = \frac{D(T_w + T_p)(1 + \alpha)}{a}$$

where:  $y$  = number of *kanban* (cards or bins)
   $D$  = demand per unit time
   $T_w$ = waiting time of *kanban*
   $T_p$ = processing time
   $a$  = container capacity (not more than 10% of daily requirement)
   $\alpha$  = policy variable (not over 10%).

Their model was developed for a deterministic environment. The "policy variable" is an attempt to compensate for production variability. In our experience this equation gives acceptable results only when demand is relatively steady and production processes are relatively stable; it is presented here for completeness rather than as an endorsement of its usage.

Hopp, Spearman, and Zhang (1997) have developed a much more sophisticated stochastic model that provides a user-established level of customer service and is optimal from a cost standpoint. It takes into account the cost of each item, average demand, the uncertainty and variability of demand during lead time, and the frequency with which management is willing to order items. Their algorithm can be used with any spreadsheet program. Their model was not specifically developed for pull production control, but we have found that it works remarkably well in a production environment. We *highly* recommend it to anyone establishing pull quantities and reorder points.

# Getting Started

So, we have our bin quantities, our reorder points, and our marketplace quantities. Where do we go from here? One way is a quick trip to the discount store where bins can often be purchased for less than a dollar. In other words, *just get started!* Schonberger seems to agree: "Kanban is something that can be installed between any successive pairs of processes in fifteen minutes, using a few containers and masking tape" (1990: 308). His point is well taken: setting up a pull system is not that tough. *Once the vision of the future has been developed*, there is really no reason to delay.

Even though the mechanics of a pull system are easy to implement, it is important to plan the system carefully. After all, the factory will be using the pull system to control its production and material flow. Hopp and Spearman observe that it is possible "to confuse simplicity of ideals with simplicity of implementation" (1996: 178). Just because a pull system is easy to put in place doesn't mean it is a simple system. Pull systems are "deceptively simple" (Wrennall and Markey 1995: 21), and we emphasize that pull production control should be part of a coherent manufacturing strategy.

Pull production control requires careful planning and an overall vision of the future. The value stream mapping technique (Chapter 6) is an excellent way to develop a vision of the factory's future state. If a pull system becomes an end in itself, a factory may generate a maddening array of cards, containers, and policies. This is distracting and inefficient. The goal is to run today's factory as efficiently as possible, not to create a needlessly complicated card system that usurps resources and renders the factory unresponsive and noncompetitive! We have seen one instance where *kanban* cards were used to communicate a production schedule that had been generated by an MRP-type (push) production control system. The management wanted to use *kanban*, so instead of sending the MRP-generated schedule to the shop floor on a computer printout, the daily production schedule was sent out to the floor as a stack of *kanban* cards.

The workers involved in pull production control have increased autonomy, so it is very important that they understand the underlying concepts, not just the mechanics, of the system. We suggest a hands-on pull system demonstration followed immediately by implementation in a particular product family. Training that takes place weeks or months ahead of time causes anxiety and false expectations, and the participants usually require a refresher session later. Training that is followed immediately by implementation seems to work well.

Several manufacturing authorities assert that certain prerequisites must be in place before pull production control can be attempted. These prerequisites include low inventory, excellent quality, reliable processes, short cycle times, short setup times, cooperative worker attitudes, uniform demand, and uniform production schedules. In fact, Nicholas (1998: 269) concedes, "It should be obvious that pull production will be a long time in coming for some organizations, and will never come for others." *We disagree vehemently! Don't be discouraged!* It is true that the prerequisites listed above would benefit any factory, but *they are not necessary or required before pull production control is instituted.* In an article entitled "Creating a Culture of Change," Marine and Riley (1995:30) make a similar observation:

> Many contemporary change programs for Total Quality Management (TQM), Just-in-Time (JIT), (SQC), and so on, advocate a checklist or organizational prerequisites that must be met to ensure success. In most cases, if those prerequisites were met, the company would have already made significant progress toward the program objectives! As organizations, we do not typically have the luxury of time and foresight to prepare so completely before declaring war! The key is to assess the organizational strengths we have and deploy them with a strategy that can gain ground while the other prerequisites are addressed.

We often hear manufacturing managers admit apologetically that pull systems won't work in their factories because the . . . setup times are too long . . . quality isn't good enough . . . machines are too unreliable . . . workers don't get along with one another . . . and customer orders are too unpredictable. Interestingly, there is nothing in these statements that precludes a pull system. In fact, these factory problems are even more severe when production is controlled by a push system. By making these "apologies," the managers are actually strengthening the case for why they *should* be implementing a pull system!

Recently we were reminiscing with a manufacturing manager who has successfully implemented a pull system. He told us that initially he thought a pull system was completely out of the question for his factory because of its many serious production problems. The articles and books he had read warned that pull systems should not be attempted until the problems are solved and the customer orders are smooth. After he discussed the subject with us, he realized that the pull system would not only solve many of his chronic production problems but would solve them immediately!

That week he ordered bins for the first parts to be pulled, but the bins did not arrive when scheduled. The following week we went with him to the local discount store and bought enough bins to accommodate four different production items. Those four parts were put on the pull system and, within days, the difference was undeniable. Stockouts of those items had completely disappeared, production was smoother, and quality had improved. Plus, inventory began to drop. In the months that followed we helped this manager add 130 more parts to the system, and the workers in his factory are adding more each week. The assembly workers support the system because they don't run short of parts, and the managers support it because their operation is much more efficient and predictable. Even the salespeople like it because the customer lead time is getting shorter every day!

A pull system is robust and self-correcting, and it is relatively insensitive to the problems and uncertainties that plague any factory. However, we must be realistic about these problems when setting up the pull system. If we hide the truth from ourselves and if we set the marketplace quantities too low, a reduction in throughput is the inevitable result. If, however, the levels are set practically and thoughtfully, the factory performance will improve in spite of other problems. Eventually, when production problems are addressed (as they should be anyway), the factory will be positioned to realize even greater benefits associated with pull production control.

## Pull Production Questions

The principles that make a pull system successful and beneficial contrast starkly with the principles of mass production and economies of scale. Because it is so different, people are naturally skeptical about pull systems. Many people have difficulty understanding how something as simple as a pull system can control the flow of material in a factory. Repeating and reiterating exactly how and why a pull system works helps reinforce workers' conditional acceptance and growing confidence in pull principles. The following common questions about pull production control illustrate the concerns and curiosity of manufacturing managers.

Q. What is meant by the term "pull?"

A. Pull is an approach to production control that is based on replenishment. Items throughout the value stream are produced only to replace what has actually been used. Usually a minimal store or "marketplace" of specific items is established. As the store is depleted, items are produced or purchased to maintain an appropriate stock level.

Q. How is a pull system different from typical production scheduling?

A. A pull system is reactionary; a traditional push system is anticipatory. A pull system reacts to consumption of material. For example, a pull signal is generated in response to material being used by a customer or a downstream process. In a push production control system, a production schedule is generated based on anticipated requirements. For example, production might be scheduled today in preparation for a current or future customer order or a pending need at a downstream process.

Q. Is a production schedule that is based on customer releases really different from pull?

A. Yes. A production schedule based on *firm releases* would be a *make-to-order system*, not a pull system. However, most releases change often (weekly, daily, and even hourly). In reality, a production schedule based on customer releases usually reflects a "forecast," and the only thing we know for sure about any forecast is that it is probably wrong.

Q. What are some advantages of pull systems?

A. Pull systems require less work-in-process (WIP) for a given throughput rate. They limit WIP because production stops when a marketplace reaches its established stock level. Consequently, inventory investment is lower. There is also less congestion on the shop floor, and floor space becomes available. The workplace can be better organized, and material can be tracked more easily. Cycle times and lead times are much shorter, and customer service improves. Pull systems are inherently easy to implement and manage on the shop floor.

Q. Why is a pull system easier to manage on the shop floor?

A. Push systems attempt to control throughput (what goes out) by regulating shop orders (what goes in). Throughput is inherently difficult to control: it can neither be seen nor counted. In contrast, pull systems control WIP, which can be seen and counted. Pull systems can be managed visually without scheduling. There are smaller quantities to track, and the marketplaces are clearly labeled and organized.

Q. Which production system is more profitable?

A. A properly running pull system will always be more profitable than an otherwise equivalent push system. Pull systems are very profitable even when the marketplace quantities have not been set at the optimal levels. Push systems are affected severely if shop orders aren't released to the shop floor at the ideal rate. If shop orders are released faster than they can be completed, WIP accumulates and costs go up. If shop orders are released too slowly, people and equipment are underutilized. In both cases, profitability of a push system declines.

Q. Can a pull system respond to changes in customer demand?
A. Yes. Pull systems moderate the effects of fluctuating customer demand.
Q. Is a pull system the best way to process material through a factory?
A. No. Today's factory needs to reduce variability wherever possible and to process material through the factory in a smooth and uninterrupted way. Ideally single-piece flow is best; it provides the shortest cycle time and lowest variability. Sometimes, however, one-piece flow is not possible. Pull systems are an excellent complement to single-piece flow.
Q. Where should a pull system be used?
A. Wherever single-piece flow is not possible. Single-piece flow is not possible if equipment is unreliable, if setup operations are long, if equipment must produce a variety of different parts, or if subsequent processes are located at great distances from each other. In these cases, single-piece flow won't work, so a pull system should be used. A controlled amount of material should be stored in a marketplace from which the downstream process can draw as needed. There are analytical techniques such as value stream mapping that can be used to determine where a pull system would be appropriate. In general, *flow where you can, and pull where you should!*
Q. Can a pull system be implemented without *kanban* cards?
A. Yes. Cards are not a requirement for a good pull system. Pull signals can be generated in a variety of ways including bins, balls, floor designations, electronics, and the Internet. Cards can be useful, however, especially when operations are separated spatially. Cards are highly visual. They can be stacked, sorted, faxed, and used for accounting purposes.
Q. How can a pull system be established?
A. First, the production process must be analyzed to develop a vision of where material should flow and from where it should be pulled. Then *just get started!* There is no reason to delay. The more formal aspects of the pull system (quantities, containers, designated spaces and locations) most likely will change sooner or later anyway; the implementation team will find more practicable quantities, more appropriate tubs, more convenient places for marketplaces. Production problems are easier to handle in a pull system than

in a push system. Even if a pull system is the only change a factory makes, the factory's performance will still improve greatly.

Q. Can a production or inventory control system always be classified as push or pull?

A. No. The concepts of push and pull have been described here in their pure forms to make the contrast clear. Most production systems exhibit elements of both push and pull and can be considered "hybrid" systems.

## Summary

A pull system is very simply a replenishment system—a way to replenish production material that has been consumed in downstream processes—it is not a zero inventory system. It is robust in the sense that it is immune to problems and fluctuations in the production stream. A pull system is also invulnerable to minor implementation errors. It outperforms an otherwise equivalent push system even when the WIP levels have been set too high or too low.

Pull systems are not magic, although managers who implement them see almost magical improvement in factory throughput, production costs, and customer lead time. Pull systems work for very sound, scientific reasons. Pull systems provide excellent protection against stockouts, and they allow high throughput rates. The WIP cap imposed by a pull system provides improved quality, less congestion, better flexibility, shorter cycle times, shorter lead times, better customer service, and lower production costs when compared to a push system. Also, pull systems are inherently easier to control.

According to Nicholas, "The charm of a pull system is its effectiveness and simplicity" (1998: 257). While we agree heartily, we also offer a caution: the simplicity is deceptive. The mechanics of a pull system are quite easy, but the implementation requires a vision of the future and careful planning. Training is important, especially since the system imparts greater autonomy to the workforce. In essence, a pull system allows the factory to be self-aware, and the workers "control" production without the intervention of a central control office.

In this chapter, we draw a rather stark contrast between push and pull. In practice, some factories operate on a loose, hybrid system. Extremely talented materials managers and production control managers often work around their dysfunctional MRP systems by using pull techniques informally. We hear comments such as "Oh, we already do that" or "We just ignore what the system tells us, and replenish our material as it gets used." When introduced as a formalization of these existing practices, the resistance to pull production control is minimized.Implementation projects go smoothly when the managers and workers involved realize that a pull system is just a simple way for today's factory to control its own production.

*Chapter 10*

# Achieving Quality at the Source

## Introduction

Today's factory is judged by the quality of its products and services. More than ever before, customers are demanding excellent quality in the goods and services they buy. Excellent quality is now considered a prerequisite for doing business rather than a competitive advantage. The importance of high quality cannot be overstated.

In the 1980s, manufacturers began to acknowledge the high costs associated with poor quality, and they began to demand excellent quality from their own internal processes and from their suppliers. The quality of goods supplied by many factories has improved markedly. Many manufacturing companies have instituted some kind of quality assurance program complete with quality manuals, functional departments, and inspection plans to assure high quality. There is increasing customer pressure to become ISO 9000 certified or, in the automotive industry, QS 9000 certified. Certification requirements are motivating companies to formalize their approaches to quality and adopt an official quality system.

Prestigious awards also highlight the increasing emphasis on quality. In 1987 under President Reagan, the U.S. government established the Malcolm Baldridge Award for quality, a distinction named for the former Secretary of Commerce who died in 1987. Its purpose is to promote quality awareness and to publicize successful quality strategies. Curiously, companies are self-nominated for the Baldridge Award; Philip Crosby advocates that customers nominate companies (Chowdhury 1998: 14). Companies are scored in seven categories: leadership, strategic planning, customer and market focus, information and analysis, human resource focus, process management, and business results (Nicholas 1998: 130). The criteria themselves are a good blueprint for a quality system.

It may seem obvious that excellent quality results in high customer satisfaction. Both researchers and manufacturers agree: total quality management (TQM) practices have a strong and positive effect on customer satisfaction (Choi and Eboch 1998: 69-70). However, it is not so obvious what effect high quality has on factory performance or corporate earnings. The anecdotal evidence is inconsistent and inconclusive, and on this point there is no general agreement among researchers or manufacturers. Recently academic researchers have turned their attention to the effect TQM practices have on factory performance and quality performance.

TQM practices contribute notably to the reduction of cycle time, which we have shown to be a very important performance measure (Chapters 4, 5, and 6). Shorter cycle time has many benefits for the factory: shorter lead time, higher customer service level (on-time delivery), faster response to changing customer requirements, lower inventory, lower production costs, higher revenue, and higher profit (Flynn, Sakakibara, and Schroeder 1995: 1329, 1353)! There is also a strong relationship between TQM practices and quality performance. Introducing TQM practices such as Statistical Process Control (SPC) and operator training leads to a reduction in scrap, rework, inspection, and the associated costs (Adam 1994).

The research proves that TQM practices are scientifically valid and that they *can* have a significant and beneficial effect on factory performance. However, the experience of the industrial community as a whole has not been entirely successful. Consider an Arthur D. Little survey of 500 executives from U.S. manufacturing and service firms. The report indicates that only one-third of respondents believe TQM helped their companies be more competitive. A similar study of 100 British firms found that only one-fifth of the TQM programs improved significantly their organizational effectiveness. The American Electronics Association found that 63% of firms reported that TQM practices failed to reduce defects by 10% or more (Choi and Eboch 1998: 62).

Why is there such a disparity between the potential and the actual benefits of TQM? One very plausible explanation is that TQM practices are often implemented at the request or insistence of a customer. This leads to the implementation of specific TQM practices that will satisfy the immediate customer requirement rather than selecting practices as part of a holistic management strategy. Piecemeal application of specific practices cannot possibly confer the tremendous competitive advantages of a systemic TQM philosophy.

Even so, many manufacturing companies are reluctant to take a system-wide approach to TQM. Perhaps management believes a TQM implementation is too expensive or that the benefits are not worth the investment. After all, the belief that excellent quality is "too expensive" is still held firmly by many manufacturers. However, there *is* a relationship between quality and corporate performance. Let's look at the corporate performance of several companies that have won the Malcolm Baldrige National Quality Award. The National Institute of Standards and Technology (NIST) has developed a stock index comprised of Malcolm Baldrige Award winners. Through 1997 this index outperformed the S&P 500 by a margin of 2.7 to 1. It gained a 394.4% return on investment compared to 146.9% for the S&P 500 (Aldred 1998a: 9). Clearly, there *is a strong correlation between quality and corporate performance.*

This chapter explores two major quality assurance tools, *statistical process control* (SPC) and *mistake-proofing*. We briefly discuss the theory behind SPC and the assumptions that limit its application and effectiveness. We explain the principles of mistake-proofing, and we discuss how mistake-proofing devices can be developed and applied. SPC and mistake-proofing have distinct but complementary functions in a quality program. They work together to provide higher quality than can be achieved by using either tool independently.

How high should a company aim when setting its quality goals? In this chapter we provide compelling evidence that supports a surprising answer: perfect quality. *Optimal performance is achieved only when quality is perfect.* This chapter presents clear strategies for achieving perfect quality, defined as zero defects, and *a new way of thinking about quality.*

# The Cost of Quality

Many American manufacturers believe that excellent quality simply costs too much. This belief is so deeply entrenched that it can be considered an underlying assumption of American manufacturing culture. Company brochures may boast of high quality levels and low ppm (parts-per-million defective) levels, but many manufacturing managers still believe it would be more profitable to operate at lower quality levels. The rationale is the mistaken assumption that achieving better quality requires a large financial investment, such as expensive equipment or tooling, additional inspection, slower production rates, and so forth. The result of this reasoning is a company policy that advocates providing products of minimal, acceptable quality to the customer or to the next production process.

Figure 10.1a is a graphical representation of this anachronistic view of the cost of quality. The curve shows that manufacturing cost is lowest when the quality level is lowest. As quality improves, the production cost rises, eventually becoming infinite as quality approaches perfection.

In the 1980s and 1990s quality gurus such as Joseph M. Juran promoted a dif-

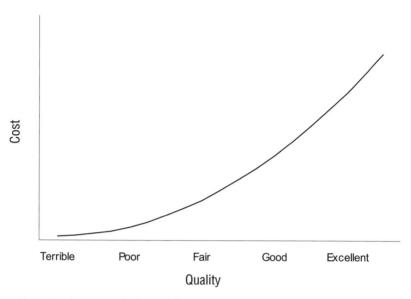

**Figure 10.1a** *The Traditional View of the Cost of Quality*

ferent view of the cost of quality (Figure 10.1b). Manufacturers began to acknowledge that poor quality has associated costs that were previously unrecognized. These costs include:

- scrap
- rework
- longer cycle time
- more elaborate inspection procedures
- loss of customer good will
- loss of reputation
- loss of market share
- disruptions caused by defects being passed along the production line.

According to this view, the manufacturing cost decreases as quality improves until the cost of preventing defects begins to outweigh the financial benefits associated with higher quality. That optimal point is where the cost of conformance exactly balances the cost of nonconformance (Love, Guo, and Irwin 1995: 403-4).

The cost of poor quality is inherently difficult to quantify. Functions such as Genichi Taguchi's loss function have been used in attempts to quantify these costs. Nevertheless, this perspective of the cost of quality (Figure 10.1b) leads to considerable disagreement in the factory.

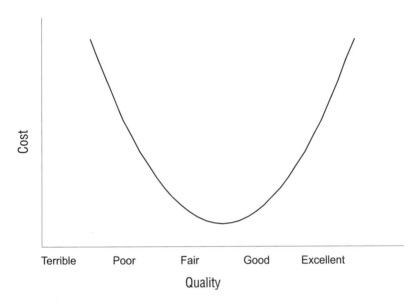

*Figure 10.1b Acknowledging the Cost of Poor Quality*

Some manufacturers are beginning to realize that the cost of poor quality is *much greater* than previously thought. Researchers such as Philip Crosby and Robert Cole have promoted a third view of the cost of quality. They maintain that production cost *continues to decrease* as product quality improves. In addition, Cole emphasizes that high quality provides substantial financial advantages, and that some customers are even willing to *pay* for higher quality (Cole 1992: 118-19). There are also intangible advantages of high quality. A factory culture that encourages the production of high-quality goods also nurtures the morale and motivation of workers and has positive company-wide impact.

Significant quality improvement can be achieved at minimal cost by looking for solutions that are creative rather than costly. Improving the quality of a product often does not require capital investment in expensive, high-tech equipment or hiring more workers or inspectors. It does require a *different way of thinking* about quality problems. It requires addressing production problems before they result in defective goods, in other words, *quality at the source*. This reduces the cost of production by eliminating scrap, rework, inspection, and other costs usually associated with poor quality.

The curve in Figure 10.1c shows the high cost associated with poor quality and the ever decreasing cost as quality improves. This means that the lowest manufacturing cost is achieved only when quality is perfect.

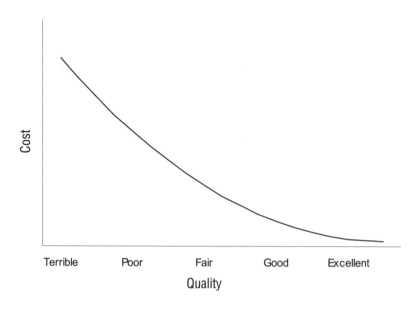

*Figure 10.1c A New Perspective on the Cost of Quality*

The views represented by Figures 10.1a–10.1c seem contradictory, and there is no clear consensus among researchers as to which view is correct. These issues continue to be debated in the quality literature. Although the perspective represented by Figure 10.1a is still widely believed by many manufacturers, most quality researchers agree that this view seriously underestimates the cost of nonconformance and is therefore not a valid model.

We offer a conciliatory perspective that reconciles Figures 10.1b and 10.1c; it is well supported by rigorous analytical research and empirical evidence (Li and Rajagopalan 1998: 1528; Love, Guo, and Irwin 1995: 415). Figure 10.1b offers an excellent *short-term* view of the cost of quality: there *is* an optimal quality level that provides acceptable quality at minimal cost. Figure 10.1c is provably optimal as a *long-term* strategy: production cost declines steadily as the factory's defect rate approaches zero. Our new perspective recognizes the high cost of nonconformance while preventing excessive expenditure on appraisal (inspection) and prevention (10.1b). We also acknowledge that production cost declines steadily as the defect rate approaches zero (10.1c). As a long-term strategy, today's factory should work continually to improve the quality of its goods and services.

## Quality Strategies

Inspection continues to be one of the most common approaches to quality assurance. There are three types of inspection: judgement inspection, informative inspection, and source inspection. Each has its own specific purpose. We will discuss the role each type of inspection can play in an overall quality assurance system.

## Judgement Inspection

The most popular type of inspection is judgement inspection. Judgement inspection distinguishes defective product from product that conforms to relevant specifications. It can involve sampling or 100% examination of the product. Acceptable quality level (AQL) sampling is an example of judgement inspection.

Even though its purpose is to assure quality, judgement inspection is never 100% effective. Nevertheless, customer requirements or extenuating circumstances sometimes require judgement inspection. For example, the aerospace industry, the U.S. government, and the petroleum industry often require some form of judgement inspection of purchased items. Also, if any defective items are accidentally produced, judgement inspection is usually used to find and remove them.

We are familiar with a valve manufacturer that produces pressure relief valves. The last stage in the manufacturing process is a staking operation that physically closes the valve, permanently preventing it from being reopened. Often the valves are assembled with internal components missing. The manufacturer devised a judgement inspection that uses sound to verify that all components have been installed properly. An operator picks up each valve and holds it near his or her ear. Then the operator shakes the valve. If the valve rattles, components are necessarily missing and the valve is considered defective. This is an example of 100% judgement inspection. Like all judgement inspections, the shake test can never be 100% effective, and this manufacturer continues to ship valves with missing components.

Judgement inspection is often ridiculed as an attempt to *sort out defects* or *inspect quality into the product*. This is because it does not improve the process, it does not reduce the number of defects produced, nor does it catch all the defects. If judgement inspection is the only effort made to assure quality, some defects will always slip through. Sometimes judgement inspection is necessary because of contractual obligations or extenuating circumstances. Nevertheless, today's factory must arm itself with more powerful quality assurance tools in order to pursue the goal of zero defects.

## Informative Inspection

Informative inspection is a more sophisticated approach to quality assurance. Unlike judgement inspection, informative inspection can be used to control, and even improve, the process. Statistical process control, successive checks (by the next process in the production sequence), and self-checks are examples of informative inspection. Like judgement inspection, informative inspection scrutinizes the output of a process, but its purpose is completely different. Rather than making judgments about the product's suitability for use, informative inspection is used to monitor and control the process.

### Statistical Process Control

Statistical process control is based on the premise that production processes always have some inherent variability. This is due to unknown and unavoidable causes that influence the process in a random way. Some of these *common* causes might be trucks passing on the highway, minor inconsistencies in tooling or raw material, lack of flexural rigidity of production equipment, or the number of pigeons on the roof. If the cumulative effect of these *chance* causes is relatively minor, the process is said to be "in control." Production processes are also influenced by *special* or *assignable* causes. These might be dimensional variations in tooling, discrepancies in the physical properties of raw material, differences in operators or work procedures, and environmental factors such as humidity or temperature. A process that is under the influence of assignable causes is said to be "out-of-control" (Montgomery 1991: 102–3).

It is important to realize that SPC makes no judgements about the fitness for use of the items being produced. Statistical process control is used to monitor a process and verify that no assignable causes are perturbing it. If a process is in control, the ordinary random variation in the process output is due only to chance. The output of an in-control process is stable and predictable. Whenever the process output varies more or less than the predicted amount, an out-of-control condition is signaled. It is important to emphasize that all out-of-control signals are noteworthy even if the change is advantageous. SPC, therefore, is not specifically designed to improve the process but to monitor the stability, consistency, and predictability of the process.

In actual applications, SPC usually involves a significant time delay between the onset of an assignable cause and the inspection of the next sample from the process output. Often samples are taken and evaluated hourly; if a process makes 100 pieces per hour, then 100 pieces are produced between samples. SPC assumes that the assignable cause continues throughout the delay; as a result, it cannot detect fleeting problems. Often the out-of-control condition is not indicated in the first sample following the special event. This means SPC can continue to indicate that a process is "in control" even *after* the onset of the special cause. So, the process is assumed to be in control even though it is no longer behaving in a stable, predictable way. Depending on the sampling frequency and the production rate, it is possible for thousands of parts to be produced by an out-of-control process before there is any indication that the process is no longer in control. This lack of sensitivity is a common topic in quality assurance literature, and techniques have been developed for improving the sensitivity of SPC. However, there is always some delay between the event and the signal, and today's factory should be aware of the costly ramifications associated with such a delay.

### Successive Inspection

Successive inspection relies on a subsequent process to check the output of a previous process. The user or "customer" is arguably the best inspector since a defect or deficiency resulting from the previous process can render an item unfit for use or further processing. Successive inspection involves 100% inspection, but its purpose is not judgement. Its purpose is to provide immediate feedback about the previous process. The feedback is much faster than with SPC, especially if there is little work-in-process (WIP) between the processes. However, successive inspection is still "after the fact."

A brief example of successive checking: an operation produces subassemblies from a variety of components. One component gets marked with an identifying ink stamp. In this process it is important that the ink stamp is clear, centered, and oriented properly on the assembled device. The machine that applies the ink stamp automatically ejects the parts through a tube into a closed container, so the operator doesn't see the output of her process before the parts are conveyed to the assembly area. When an assembly operator orients the marked component, she also checks the ink stamp to be sure it is clear and centered. If identification marks are blurry or off center, the component is rejected, and the stamping process is evaluated.

### Self-Inspection

Self-inspection is performed immediately after the process is complete. It is still "after the fact," but it has some significant advantages over the other forms of inspection we have discussed. It gives the fastest possible feedback about the previous process; and it catches the *first defect*, so corrections can be made before more defective items are produced.

Informative inspection is tied directly to the process either by sampling the process output (SPC), by checking the process output at the next process (successive), or by checking the process output immediately after the process has been performed (self). It provides faster feedback than judgement inspection, so fewer defects are made and the causes can be identified more easily. Of the three types of informative inspection, SPC provides the slowest feedback, and self-inspection provides the fastest since the very first defect is caught.

It might seem that informative inspection is similar to closing the barn door after a horse has already run out; in fact, that analogy is ideal for describing *all* types of inspection. Self-inspection is *exactly* like closing the barn door after a horse has run out. This prevents other horses from running out and provides a chance to keep the first horse from running very far. It is obviously less than ideal, however, since one horse does escape.

Successive inspection is analogous to having your neighbor call on the telephone to say, "Hey, Dale, you better close your barn door. Your horse just ran through my garden!" Closing the barn door in response to this telephone call means that no more horses can run away. However, several other horses might already have escaped from the barn depending on the time delay between when the first horse ran out and when the neighbor saw it in the garden.

SPC can even be represented using the horse–barn analogy. Every night, choose five stalls at random, check to see whether there is a horse in each stall, plot the results on a control chart, and monitor the chart for out-of-control signals. In this way we can "control" the process so that not more than a previously specified number of horses runs out during any specific day. Again, this approach is perfectly valid, but it necessitates some horses running out before an out-of-control signal is generated. Also, the presence or absence of horses in the stalls chosen for inspection does not necessarily indicate anything about other horses from other stalls.

Judgement inspection can also be illustrated using the horse–barn analogy. Judgement inspection advocates checking all the stalls every night to see whether any horses ran out during the day. While this approach certainly gives the most accurate picture of how many horses are in the barn on any given night, it isn't very useful because of the probable long delay between a horse running out and the evening count.

### *Source Inspection*

Source inspection is a completely different approach to inspection. All other forms of inspection monitor the output of a process, but source inspection monitors the process itself to ensure that it is being performed properly. Therefore, source inspection takes place "before the fact," and an "out-of-control" signal is generated before any defective items are produced. This is why source inspection is preferred above any other type of inspection. Mistake-proofing, which is an important tool for achieving "zero defects," is closely related to source inspection.

In the horse–barn analogy, source inspection is checking the status of the door, not the status of the horses. If the door is open, a signal is generated, and someone closes the door. This precludes the possibility of a horse running away because the opportunity (the open door) is corrected immediately or never occurs.

# The Importance of Early Detection

Fast feedback about a process is important for two reasons. First, the cost associated with a problem increases the longer the detection of the problem is delayed. If a problem or defect is detected immediately where it is caused, the disruption and cost are minimized. Detection at a subsequent process is slightly more disruptive and also more costly. Some defective items are in the system, and they must be scrapped or repaired. These costs increase dramatically if detection is delayed until final inspection. By then, a significant number of defective items are in the system, and rework is much more extensive and complicated. It may happen that the defective items cannot be repaired, and the entire order needs to be rescheduled. This causes considerable disruption to the factory. It is even worse if defective items are shipped to the customer because then the factory incurs substantial cost. This is due to the high cost associated with returns, recalls, warranty obligations, liability or litigation, loss of good will, loss of reputation, and perhaps loss of market share.

The second reason fast feedback is important is because the ability to determine root causes diminishes with time. As the time lag between an event and its discovery is extended, identification of the true cause of the event becomes more difficult. To prevent future defects, root causes must be identified and eliminated. Solving tertiary problems or masking symptoms may improve quality in the short run and may serve as an emergency measure, but the underlying problem must be solved if today's factory is going to achieve the goal of zero defects!

# From SPC to Zero Defects

It is easy to understand the allure and popularity of statistical quality control (SQC). It appeals to our desire to manage our factories objectively and scientifically rather than by perceptions and opinions. The reliance on systematic data collection, statistical methods, charts, and graphs lends an aura of rationality to the often chaotic manufacturing environment. Unfortunately, this scientific approach to managing quality also has some hidden side effects that can deter today's factory from achieving or even pursuing the goal of zero defects. We have chosen four of these side effects for discussion.

1. **Workers are distanced from the quality of their product.**
   Statistical process control is based on mathematics. The calculations themselves are not particularly complicated, but the underlying statistical theory is somewhat sophisticated. Not all shop floor workers have adequate mathematical background or adequate training to understand fully the statistical tools they are expected to use. Often a worker is asked to plot data

on an SPC control chart for a quality professional to interpret later. This definitely distances the worker from the product and its quality, and it diminishes the worker's sense of ownership in the process.

2. **Managers and executives are disengaged from quality improvement.**
   Elsewhere in the factory hierarchy, similar side effects are taking place. Production managers and corporate executives can become disengaged from the quality improvement effort if responsibility is relegated to a specialty department staffed by statisticians. Consequently, the shop floor workers, who are responsible for making the product, and the management, the people responsible for running the factory, are no longer directly involved in pursuing perfect quality. Moreover, the people who are responsible for quality are often removed from the shop floor, and they are usually not included in strategic planning or corporate decision making. The specialization fostered by a factory's reliance on SPC can work against the cooperative pursuit of quality.

3. **The means (SPC) is often confused with the end (perfect quality).**
   Also, there is a danger of confusing the means with the end. During the 1980s, the growing popularity of SPC led to a proliferation of control charts that hung like wallpaper in many factories. Dennis Pawley, a former Vice President of Chrysler, remarked at the 1997 University of Michigan Lean Manufacturing Conference that the onslaught of control charts in the 1980s did save Chrysler money. The walls didn't need to be painted because they were covered with control charts! Charts and graphs are merely tools of quality improvement; they are not goals in themselves. The goal is to continually strive toward perfect quality.

4. **Reliance on SPC can discourage continual improvement.**
   The terminology, objectivity, and visibility of SPC provide a strong sense of "control." A shop floor supervisor might ask, "If a process is already under control, why should it be changed?" Having scientific evidence of stability and predictability makes it difficult for people to accept the idea that improvement is necessary. When a component is causing difficulty in an assembly operation, we often hear comments such as "it fits the gauge," "the chart shows the process is in control," and "that's not a critical dimension." While these comments may be true, they reflect an attitude that is not conducive to continual improvement. The pursuit of zero defects or perfect quality requires an environment of continual change. *Situations that were considered good enough in the past are not good enough now if they can be improved.*

# SPC Control Chart Cautions

One of the most common SPC tools is the control chart. It was developed in the 1920s by Walter Shewhart and is now used by many manufacturers throughout the world. Given its wide usage, it is surprising that this tool is so often misapplied, misunderstood, and misinterpreted. The theory and mechanics of control chart construction are beyond the scope of this book, but some cautions and clarifications are especially relevant to our pursuit of perfect quality.

DeVor, Chang, and Sutherland (1992: 200) comment that "many companies waste large amounts of money by sampling so infrequently that the resulting charts have no value." Sampling should be done often enough to provide a high degree of confidence but not so often that it interferes with production or costs too much. Both the sampling frequency and the sample size affect the control chart's ability to detect out-of-control situations. Small shifts are detected more effectively by large samples. In fact, from a purely scientific perspective, sensitivity is greatest when large samples are taken frequently; however, this can be impractical financially. Quality managers must weigh sensitivity (sample size and frequency) against effort and cost when designing control schemes.

Another caution about sampling frequency is the danger of sampling *too* frequently. Most control chart schemes assume that the data are independent and identically distributed (iid). When the process output is sampled too often or when every piece is inspected, the data often exhibit serial correlation or *autocorrelation*. This means that current data are correlated with earlier data. Autocorrelation violates the underlying assumptions of SPC. Ironically, it is present in more than 70% of industrial situations (Alwan 1991; Standard 1997).

Autocorrelation tends to occur in processes that are sensitive to chemical composition or temperature changes. It also occurs in processes that are subject to tool wear such as machining, stamping, and molding. For example, the parts produced during an extended stamping run may differ dimensionally due to wear on the die, and the effect on any one part produced is related to the effect on the parts produced before and after. When autocorrelation is present, traditional control charts are invalid (Alwan 1991; Montgomery and Mastrangelo 1991)! Fortunately, other control charting schemes have been developed to control autocorrelated processes (Standard 1997). SPC users should be aware of these complications when choosing their control schemes and sampling frequencies.

Two other features of control charts that are often ignored are *producers' risk* (alpha risk) and *consumers' risk* (beta risk). Simply put, they estimate the likelihood that the SPC control scheme will misinterpret the data. Producers' risk is the probability of concluding that a process is *out of control* when it is actually *in control*. Conversely, consumers' risk is the probability of concluding that a process is *in control* when, in fact, it is being perturbed by some assignable cause. These risks are influenced by control chart parameters such as sample size and control limits. The standard 6-*sigma* control chart commonly used in manufacturing today has a producer's risk of 0.0027. This means that there is a 0.27% chance that a data point will fall outside the control limits even though the process is in control. This is sometimes called a *false alarm*.

A closely related feature of control charts is the *average run length* (ARL). The ARL is the expected number of samples taken until there is an out-of-control signal. Control schemes have an in-control run length and an out-of-control run length. Ideally the in-control run length would be infinite, meaning the process would never generate a false alarm. Conversely, the ideal out-of-control run length would be 1.0. This means there would be a high likelihood of detecting the out-of-control situation on the first sample following the onset of the assignable cause.

Consider a process that machines a part to a specific diameter. The process mean is 1.000 inch and the standard deviation of the process is 0.25 inch. Now, let's say that the process goes out of control and the diameter of the parts increases by 0.125 inch; the process mean is now 1.125 inches. For convenience, we can represent this increase as a multiple of the process standard deviation. In this case, the process mean has "shifted" by 0.5 standard deviations. Is it likely that the next sample taken will indicate the process has gone out of control? No. In fact, there is less than a 5% chance that the control chart will reflect or detect this shift in the next sample.

Table 10.1 lists a variety of mean shifts, the probability of detecting the shift on any given sample, and the average run length. The sample size is 5 pieces, and standard 6-*sigma* limits are assumed. As Table 10.1 indicates, the standard 6-*sigma* control chart is not particularly sensitive to small shifts in the process mean. Large shifts, however, are quite likely to be detected in the first sample after the shift occurs.

| Sample Size | Size of Mean Shift | Probability of Detecting Shift on Next Sample | Average Run Length (out-of-control) |
|---|---|---|---|
| 5 | 0.5 | 0.03 | 33.4 |
| 5 | 1.0 | 0.22 | 4.5 |
| 5 | 1.5 | 0.64 | 1.6 |
| 5 | 2.0 | 0.93 | 1.1 |
| 5 | 2.5 | 1.00 | 1.0 |
| 5 | 3.0 | 1.00 | 1.0 |

*Table 10.1* Control Chart Effectiveness for a Variety of Mean Shifts

Let's assume that a process is "controlled" by a standard 6-sigma control chart. An assignable cause is perturbing the process causing the process mean to increase by one standard deviation. According to Table 10.1, the probability of detecting the shift is 22% in any given sample, and the "out-of-control" run length is 4.5. This means an average of 5 samples is required before the out-of-control condition is detected. If samples were taken every hour, which is customary in many factories, it would take on average 5 hours to detect the problem. So, if the process were producing 1 piece every 18 seconds, *1000 pieces* would be produced before the control chart alarms us that the process is no longer stable! This delay has an extremely high cost: the process continues to generate product that may be defective, and the ability to find the root cause diminishes with every passing moment.

The last caution we present concerns process capability. Process capability reflects the ability of a process to produce parts that conform to product specifications. Many original equipment manufacturers (OEMs) insist that their suppliers demonstrate excellent process capability. Process capability can be expressed mathematically by measures such as the process capability ratio (PCR), $C_p$, $C_{pk}$, and defective parts per million (ppm). The measures differ in their precise mathematical definition, but the intent is the same: to verify that the probability of producing a defective item is below a predetermined, acceptable limit.

This limit is often expressed in terms of ppm. For example, "6-*sigma*" process capability (+/- 3-sigma) is equivalent to 2700 defects per million, or 2700 ppm. Unfortunately, most factories operate far below this level of quality and produce far more than this number of defects.

"Most manufacturers in the United States operate at about 3-sigma, churning out 66,000 bad parts for every million produced. These companies lose up to 25% of their total revenue due to defects" (Murphy 1998). Murphy uses one-sided specification limits in his estimate of defect rate. If the specification has both a minimum and a maximum, 3-sigma process (+/- 1.5 sigma) capability corresponds to nearly 134,000 defects in every million parts produced!

There is confusion regarding how process capability should be calculated. Consider a supplier who must demonstrate a process capability ratio (PCR) of 1.3 to satisfy customer or contractual agreements. (PCR is defined as the total width of the tolerance band, the upper specification limit minus the lower specification limit, divided by six times the process standard deviation.) A process with a PCR of 1.3 is capable of producing only 60 defects per million.

$$PCR = \frac{USL - LSL}{6\sigma}$$

Usually the process standard deviation is not known, so an *estimate* of true process capability must be calculated after sampling the process output. Let's assume that if the actual PCR is less than 1.3, the supplier would like to detect this deficiency with a high degree of confidence, for example, 90%. Similarly, if the process capability is truly excellent, say a PCR of 1.6, the supplier would like to conclude with the same high degree of confidence (90%) that the process is *capable.*

How many pieces should the supplier inspect to be confident that the process has the required 1.3 PCR? What is the minimum PCR the supplier must *observe* in the sample to be confident that the *true* process capability ratio is at least 1.3? These questions are rarely asked in industry, but they should be. Kane (1986) and Montgomery (1991) present analytical techniques for answering them. Even though the details are beyond the scope of this book, we present the answers to illustrate our point.

*Eighty pieces* need to be inspected, and the observed PCR in the sample must be at least 1.43 to provide the desired level of confidence! Simply observing a 1.3 PCR in the sample or inspecting fewer pieces would not provide a high enough level of confidence that the process is truly capable. This example highlights that some common quality assurance practices may be questionable from a rigorous scientific standpoint.

These cautions about SPC are not intended to dissuade our readers from using this powerful tool. However, we do emphasize that SPC is not a panacea and that it is susceptible to misuse. As a stand-alone practice, SPC is an insufficient tool for achieving perfect quality. *When SPC is used in conjunction with other quality tools, zero defects can be achieved.* As we will see, mistake-proofing is a highly effective and complementary quality assurance tool.

## Putting Defect Levels into Perspective

To put defect levels into perspective, we have compiled a few examples from everyday life based on figures from the University of Michigan and the U.S. Bureau of the Census. If 99% quality were "good enough," then the University of Michigan Hospital would give the wrong baby to the new parents 33 times each year, 19.2 million pieces of U.S. mail would be delivered to the wrong address each week, and the electricity would be off 14 minutes every day. If 99.9% quality were "good enough," 117 miles of highway in Michigan would be impassable, more than half a million U.S. airline passengers would lose their luggage annually, and nearly 200,000 inedible chickens would be sold each year!

As consumers, we clearly do not tolerate such terrible quality; yet, we ask our customers to accept it every day. Discriminating OEMs are demanding more from their suppliers, and some manufacturers are taking the challenge seriously. During an 18-month period from 1994 to 1995, General Motors' Ramos Arizpe engine plant in Mexico shipped 1.5 million hydraulic valve lifters to Toyota in Japan with a zero defect rate (Moskal 1995: 33). The world-class quality standard is now approximately 2 or 3 ppm (Murphy 1998).

# The Theory of Mistake-Proofing

Pursuing perfect quality requires that we understand why defects occur so we can eliminate their causes. A *defect* is a deviation from a specification, for example, an item that does not conform to its design, material, or customer requirements. There are three major sources of manufacturing defects: excessive variance, mistakes, and complexity. Each source of defects has its own set of solutions and its own methods of control. For example, excessive variance is monitored and "controlled" through statistical methods. Mistakes are prevented through mistake-proofing. Complexity is addressed through simplification schemes such as robust design, design for manufacturability, and concurrent engineering. Complexity can lead to perplexing manufacturing problems; it also exacerbates the effects of excessive variance and mistakes.

Because this book focuses on manufacturing, we will address the two factors that affect quality and are also directly controlled by today's factory: excessive variance and mistakes. (Readers who would like to know more about the third cause of defects, product complexity, are directed to one of the excellent books on the subject of concurrent engineering such as *Concurrent Engineering Effectiveness: Integrating Product Development Across Organizations* (Fleischer and Liker 1997).)

A *mistake* is a deviation from an intended process. Human error, malfunctioning machines, or improper environmental conditions are examples of mistakes that can lead to a defect. A mistake may consist of performing a prohibited action, not performing a required action, or performing a required action incorrectly.

There is no reason to expect the familiar statistical techniques used for controlling variance to have any effect on mistakes. The occurrence of mistakes is not even well modeled by statistics. For example, the probability distribution of screw torque cannot predict whether or not the screw has been installed. The outcome of a mistake can vary hundreds of standard deviations from the "process mean." So, if defects are caused by mistakes, *perfect* quality cannot be achieved by SPC alone, and some other preventive strategy must be introduced.

Why do we consumers reject or return items we have purchased? We get home and find that the item:

- doesn't work
- is mislabeled
- is broken
- has loose parts
- has parts missing
- isn't what we ordered
- is the wrong color

These are known as *attribute defects*. An attribute has no degree of correctness: it is a feature that is either right or wrong, works or doesn't work, or is present or absent. Most customer rejections are due to an incorrect product attribute.

A *variable* is different from an attribute. It can be measured on a continuous scale. Thickness, length, hardness, chemical composition, and diameter are all examples of variable features. While attribute defects are responsible for most rejections, most quality assurance efforts focus on controlling variables.

Fortunately attribute defects are easy to catch. They are usually blunders rather than near misses. They make us say "oops," "doh," or "yikes!" Detecting attribute defects is far simpler and less expensive than detecting variable defects. For example, the presence or absence of a pin can be detected easily by passing the part under a proximity sensor. If the sensor detects the presence of the pin, the part is allowed to pass. If not, the part is directed to a rework bin. This is much simpler than trying to measure the diameter of every pin or monitoring the pin statistically and charting the results to determine whether the dimension falls within an acceptable range.

It is useful to categorize manufacturing defects because a single solution often can solve all the defects in a given category. In the example of the pressure relief valve described earlier in this chapter, there are three possible defects: a missing ball, a missing spring, and a missing piston. Each defect has its own separate cause. If each kind of defect is investigated independently, there could be three separate projects conducted by three separate teams arriving at three separate solutions. On the other hand, these defects belong to the same defect category, the "missing parts" category. Approaching the problem from this perspective, there may be a single solution that prevents any *defects due to missing parts*. Perhaps a scale could be placed in the production line immediately before the staking operation. If any valves were lighter than expected, we could reasonably conclude that a component must be missing, and the staking operation could be halted.

It is also useful to categorize manufacturing mistakes. Again, a single solution can often prevent an entire category of mistakes. We have prepared a brief list of common categories of defects and mistakes.

**Defects**
- Damaged parts
- Missing parts
- Reversed parts
- Wrong parts
- Wrong size or shape
- Wrong chemical properties
- Wrong physical properties

**Mistakes**
- Omitting a part
- Omitting a process
- Performing a process incorrectly
- Performing a process on the wrong part
- Performing a process in improper environmental conditions
- Performing the wrong process

**Defects**
- Wrong item shipped
- Wrong label
- Wrong package
- Damaged finished product

**Mistakes**
- Setting up equipment incorrectly
- Setting up a work piece incorrectly
- Adjusting equipment incorrectly
- Tooling set up incorrectly
- Tooling adjusted incorrectly
- Installing wrong part
- Installing part incorrectly

Are these mistakes due to forgetfulness, misunderstanding, carelessness, misidentification, lack of training, unclear instructions, lack of skill, accidents, surprise, intention, or fatigue? Often manufacturing mistakes *are* blamed on "human error," when it is the manufacturing system that allows the mistakes to happen. Admonitions such as "pay more attention" cannot prevent mistakes. Instead, the manufacturing system should be designed to disallow mistakes. It is the system that needs correction, not the individual!

Unfortunately there are situations in our factories that not only allow mistakes to occur but even provoke them. These are process "accidents waiting to happen." Sometimes these conditions are called *red flag conditions* or *red light conditions*. They are situations that allow or are conducive to mistakes and resultant defects. A partial list of these conditions is included here.

## Common Red Flag Situations

**Product and Process Design Issues**
- Adjustments
- Tooling that must be changed
- Equipment that must be set up
- Needlessly tight tolerances
- Many or mixed parts
- Symmetrical parts
- Asymmetrical parts
- Rapid repetition of process steps
- Long or complicated work procedure

**Management Issues**
- Unclear procedures
- Vague specifications
- Infrequent production
- Poor lighting
- Poor housekeeping
- Debris or foreign matter
- Distractions

*Mistake-proofing* is the science of preventing defects due to mistakes in the production process. Let's develop an *improvement strategy* from a mistake-proofing perspective.

- Inspection will not prevent a defect, so detect the defect when it is made.

- Inspection will not prevent a defect, so detect the defect when it is made.
- A defect is caused by a mistake, so detect the mistake when it occurs.
- A mistake is provoked by an improper condition, so detect the improper condition and prevent the mistake.
- If a defect is produced, prevent the defect from passing to subsequent processes by early detection and immediate corrective action. By monitoring those conditions that give rise to mistakes, it is possible to eliminate mistakes altogether. This is the fundamental philosophy of mistake-proofing.

Murphy's Law:                If it can go wrong, it will.
Mistake-Proofing Corollary:  If it can't go wrong, it won't.

In Figure 10.2 the following mistakes might be made in the manufacturing process:

- forgetting to drill the holes in the lid or the base
- forgetting to tap the holes in the base
- forgetting to install or tighten the screws
- installing the lid backwards
- painting the box the wrong color.

Even if all the components were dimensionally correct, some boxes would still

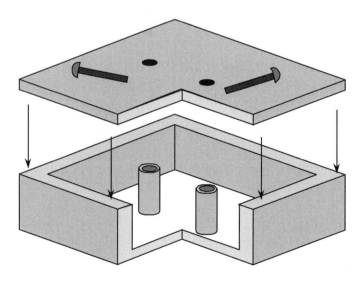

*Figure 10.2 A Simple Assembly (Adapted from Hinkley and Barkan, 1995:243)*

be defective because of assembly mistakes. How many boxes would be defective? We can compute easily the expected number of defective boxes using some very straightforward mathematics.

Rook determined in 1962 that one operation in 33,000 will be skipped and not caught or detected (Hinckley and Barkan 1995: 243-4). This corresponds to approximately 30 ppm. The cumulative effect of the potential assembly mistakes can be determined easily by computing the probability of not making *any* assembly mistakes. If we have one chance out of 33,000 of making a mistake that doesn't get caught, we must necessarily have 32,999 chances out of 33,000 of *not* making a mistake. So we have a 99.99697% chance of performing any single manufacturing step without a mistake. Let's see *how many mistakes are possible* in manufacturing the box in Figure 10.2: drilling holes (4 mistakes), tapping holes (2 mistakes), installing screws (2 mistakes), tightening screws (2 mistakes), installing lid (1 mistake), and painting (1 mistake).

In order to *produce a defect-free box*, we must perform all the manufacturing steps correctly. How do we compute the probability of making a box without any mistakes? We simply take the probability of performing any single step correctly and raise it to a power equal to the number of steps in the process. In this case, we raise 0.9999697 to the 12th power. So, we have a 99.96364% chance of making no mistakes while assembling the box. Although this seems like excellent quality, it corresponds to 364 ppm, which is 100 times higher than the current world-class standard! This example demonstrates how acceptable quality standards can be exceeded during the assembly of a simple box. Imagine how easily those standards can be exceeded when a more complicated product is manufactured!

Can zero defects ever be a reality? The obvious answer might seem to be "no." We have seen that even when components are perfect, assembly mistakes contribute approximately 30 ppm. Yet, *world-class companies are achieving zero defects* for some products for periods exceeding one year! How is this possible? Mistake-proofing is a strategy that can reduce defects far below the level attainable using SPC alone. This led John Grout to make the following prediction at the 1995 ASQC Texas Quality Expo: "If defects = 0 is possible, mistake-proofing will be in the toolbox of those who achieve it."

# Mistake-Proofing Devices

Mistake-proofing devices provide quick, efficient, 100% inspection of products and processes. They detect defects that have been made and prevent their being processed further. They can also detect mistakes that have occurred and identify conditions that allow mistakes to occur. The inspection that mistake-proofing devices provide is different fundamentally from judgement inspection. Mistake-proofing devices provide timely information about the process. If mistake-proofing is used simply as a filter to screen out defective product, the delivered quality might improve, but the huge potential benefits of mistake-proofing will go unrealized. To take full advantage of mistake-proofing, there must be a response mechanism that addresses the problems immediately. In our horse–barn analogy, a warning signal that indicates the door is open is useful only if someone responds by closing the door.

Mistake-proofing devices can be used for many purposes. They detect abnormalities, notify people, and initiate corrective action such as stopping production and capturing defects. They promote the discovery of root causes by providing instant feedback about the process. Mistake-proofing devices completely prevent recurrences, force correct procedures, and eliminate wrong choices. They also prevent damage to product or equipment and prevent personal injury.

We can use the horse–barn analogy one more time to describe different mistake-proofing devices. One device might close the corral gate after a horse runs out of the barn to keep the horse from leaving the yard. Another device might sense that a horse is running toward the barn door, and it closes the door before the horse reaches it. Still another mistake-proofing device might close the door automatically thereby preventing a horse from running out. The ultimate mistake-proofing device might be a door for stable hands to use that is too small for horses to fit through!

There are three approaches to developing mistake-proofing devices. The first involves the product, and the other two involve the process. Each approach addresses a different question.

1. **Are the key item characteristics correct?**

The *item characteristic approach* evaluates the correctness of product features such as shape, weight, dimension, and color. A fruit sorter is an example of this approach. The item characteristic approach can be used for self-checks or successive checks.

How many of us remember the childhood joy of opening a brand new box of crayons and looking at all the beautiful colors? The wonder and excitement would turn to disappointment, however, if any shade were missing, even "Burnt Sienna." Binney and Smith, maker of Crayola crayons, is well aware of the aggravation that can be caused by missing crayons, and they have taken a remarkable stance. Light sensors detect the presence of all crayons in a box, and production stops if a box is incomplete. This is preferable to having customers buy incomplete boxes of crayons. The administrative costs associated with returns and the loss of customer good will would far exceed the minor cost of stopping the process and fixing the problem (Grout 1997: 35).

Tuttle Spring Design in southeastern Michigan is a manufacturer of high-precision springs for the aerospace, automotive, and furniture industries. Several of the machines are equipped with automatic measuring devices that verify in real time critical part dimensions as the springs are being manufactured. If the dimensions are within the control limits, the springs are allowed to pass to the next process. If the dimensions indicate that the process is no longer in control, the springs are diverted off line. Some of the machines at Tuttle are even equipped with feedback loops. If the dimensions begin to drift (perhaps as a result of tool wear), the machine compensates automatically by adjusting the tooling. In this way, defective items are never produced, and the customers receive perfect quality.

2.  **Have the required operations been performed?**

The *required operations approach* verifies that crucial steps have been performed. This verification can be done during the process or between processes. An example is an automobile ignition that prevents the key from turning unless the car is in park and the wheel is straight. Also, the car will not shift out of park unless the brake is depressed. Another common example is the airplane lavatory. The lights come on only when the door is locked. This prevents embarrassment and injury due to knocking the passenger in the head with the door (Chase and Stewart 1994)!

3.  **Is the process being performed properly?**

The *performance approach* verifies that a process is being performed properly. Mistake-proofing devices that use this approach include counting, remainder methods, and critical condition indicators. An electrical circuit breaker is a common household example. If too much current is drawn, the circuit breaker "trips," turning off the power.

# Examples of Mistake-Proofing Devices

The following are common types of mistake-proofing devices.

- Guides (pins, slots, or rails) are used to orient a workpiece. Use fixtures rather than hands to improve safety and precision. Use settings rather than operator-controlled adjustments. Guides and settings standardize the operation and ensure proper alignment for mistake-free processing.
- Templates (paper, plastic, wood, or metal) ensure accurate positioning. These can be used for both production and inspection.
- Limit switches or positive stops prevent over-travel mistakes and ensure proper

positioning.
- Counters or timers ensure the proper quantity, amount, and duration.
- The remainder method is similar to counting, but it is based on discharging the exact number of parts needed for an assembly operation. If any parts remain, an assembly mistake has been made.
- Sequence restriction is excellent for ensuring that the process is performed in the correct order.
- An intelligent delivery chute prevents defective material from being processed further. It sends good product to the next process and defective product to the scrap bin. (This mistake-proofing device should be mistake proofed itself since its failure mode should cause all products to fall into the scrap bin.)
- Gates stop material from continuing or prevent an operation from occurring unless certain conditions are met.
- Sensors, such as light curtains, weight sensors, vibration sensors, force gauges, pressure transducers, thermocouples, and proximity switches, can also be used.
- Critical-condition indicators check whether conditions are appropriate for defect-free production.
- Redesign either the product or process for robustness. If orientation is critical, asymmetry can be designed into the parts and fixtures; if orientation is irrelevant, then symmetry of design prevents defects.

# A Mistake-Proofing Project Plan

A mistake-proofing implementation guide can be found below. With minor adaptations, this list can be used to develop and implement mistake-proofing devices in virtually any manufacturing environment.

### Mistake-Proofing Implementation
1. Determine and define the objectives of the mistake-proofing project. Objectives might include improved quality, lower production cost, improved customer service, or fewer difficulties in downstream operations.
2. Identify and describe the defect(s) to be eliminated.
3. Understand the *type* of defect and look for similarities to past problems. Remember that a single solution can often be applied to an entire family of problems.
4. Determine where the defect was discovered.
5. Determine where the defect was made.
6. Understand the details of the current standard procedure.
7. Identify any deviations from the standard procedure.

8. Identify and *categorize* the mistake that caused the defect.
9. Understand the conditions that allowed the mistake to occur.
10. Decide what level of mistake-proofing is most feasible: detect the defect, detect the mistake, or prevent the mistake.
11. Determine the type of mistake-proofing device that is required.
12. Create the device and test for effectiveness.
13. Install the mistake-proofing device(s).
14. Measure and document the results.
15. Verify that the device meets the original objectives.

# Summary

Customers are now demanding unprecedented levels of quality. The "World Class" quality standard is approaching *zero defects*. Excellent quality is no longer a competitive advantage but a prerequisite for doing business. Consequently, manufacturers must control their internal quality and the quality of goods delivered to their customers.

Fortunately, excellent quality does have financial advantages. The traditional view that manufacturing cost increases as quality improves is no longer considered valid by industrial researchers. Empirical evidence and rigorous scientific analysis affirm that manufacturing cost continues to decline as quality approaches perfection (or as the defect rate approaches zero).

In today's factory, inspection has three distinct purposes: determine fitness for use, evaluate the process by inspecting the product, and monitor the process itself. Timely feedback and early detection are serious considerations for two reasons: first, the cost and disruption are minimized when defects are detected immediately; and, second, the ability to find the root cause of the defect diminishes with time.

Judgement inspection determines an item's fitness for use. Judgement inspection can never lead to zero defects, but sometimes judgement inspection is necessary as an emergency measure or to satisfy customer specifications and requirements. Judgement inspection offers little information about the process because there is usually a significant time delay between when the items are made and when they are inspected.

Informative inspection evaluates a process by inspecting the output of that process. Informative inspection includes SPC, successive checks, and self-checks. Informative inspection provides faster feedback than judgement inspection: the fastest feedback is through self-checking, and the slowest feedback is through SPC. The timeliness of successive checking can be affected greatly by how much work-in-process (WIP) is in between the processes.

Source inspection actually monitors directly the process rather than the output of the process. This is very different from judgement and informative inspections. Source inspection is "before the fact." It detects when a mistake has occurred or is likely to occur, so the process can be corrected before any defects are made at all.

Even though SPC is an important quality assurance tool, it has some adverse effects that can hinder a factory's pursuit of zero defects. Workers can be distanced from the quality of their product. Managers and executives can be disengaged from quality improvement. The means (SPC) is often confused with the end (perfect quality). Reliance on SPC can discourage continual improvement.

We offer several other cautions about statistical quality control. Be careful when choosing sample size and frequency. Many manufacturers sample so rarely that the resulting charts are of little value. Conversely, sampling too frequently can lead to violations of the underlying statistical assumptions on which the control chart is based. Autocorrelation in process data violates these fundamental assumptions, and, ironically, it is present in most industrial processes. The insensitivity of traditional SPC charts is reflected in their inability to catch fleeting problems and the length of time needed to detect small shifts in the process mean. Process capability, which is an important measure for today's factory, should be calculated using valid statistical techniques.

Mistake-proofing acknowledges that mistakes come from three causes: excessive variance, mistakes, and complexity. Variance is addressed through SPC, and mistakes are prevented through mistake-proofing. While SPC detects the presence of special causes, it is mistake-proofing that prevents these special causes from recurring or occurring in the first place. Mistake-proofing is a tool for quality assurance that complements SPC. Mistake-proofing is not a stand-alone practice, but without it zero defects can never be a reality.

What are the quality assurance options for today's factory?

A) Sort out defective items and keep them from "escaping."
B) Let the customer return defective products.
C) Catch defective items immediately and repair or scrap them.
D) Eliminate the cause of defects.
E) All/none of the above

Which option is most desirable for today's factory? The only option that moves the factory toward zero defects is . . . to eliminate the cause of defects (D)!

Mistake-proofing complements other quality assurance tools to completely eliminate defects. Through mistake-proofing, today's factory can achieve "perfect quality," which is provably optimal from a cost standpoint. Mistake-proofing lowers production cost, shortens lead time, improves customer service, and makes the workplace safer. Other benefits of mistake-proofing include lower training costs, lower required skill levels, higher production efficiency, and higher corporate profit.

We invite our readers to reflect on their own experience by asking a compelling question: "How does my factory run?"

| | | |
|---|---|---|
| ✔ TQM philosophy<br>✔ Built-in quality<br>✔ In-system process control<br>✔ Plant-wide nervous system<br>✔ Production stops while<br>    problems are resolved | Or | ✔ I am an island<br>✔ Sort out defects<br>✔ Statistical process control<br>✔ Hide status of plant from<br>    management<br>✔ Keep production rolling<br>    along – someone else<br>    will fix it |

*Chapter 11*

# Quick Setup and Small Lot Production

## Introduction

Quick setup is one of the most widely accepted and least controversial lean practices. Companies that embark on a lean transformation often start by reducing their setup times. We believe this is because there is an easy fit between mass production and lean manufacturing on the issue of setup time. Time-consuming setup operations cause delays that are very obvious and apparent. These lengthy operations disrupt production, and they can be aggravating to managers and shop floor workers. Whether the factory is operating under mass production tenets or lean manufacturing philosophies, most manufacturing managers agree: setup time should be reduced as much as possible.

Why do factory managers want to reduce their setup time? Surprisingly, there is no general consensus. Most factory managers can agree that setup time should be reduced, but few can agree on why. Many managers are pursuing quick setup with no clear picture of how the newly available time will be used.

Often quick setup is pursued for its own merits, in other words, to prepare production equipment more quickly for the next job. Consider a manager who sees a variance on her weekly standard labor report. She may believe that the labor variance is due to the time and difficulty associated with setup operations; if less time were spent on setup activity, more time could be spent producing parts. In fact, she is correct. If long setup time were the major contributor to the labor variance, the weekly reports would probably show improvement when setup time was reduced. This is a classic example of the mass production economies-of-scale principle.

The most commonly recognized benefits of quick setup are greater machine capacity and higher workstation efficiency. Achieving these benefits, however, depends on other factors as well. For example, if production equipment is highly utilized, there is little time available for setup operations. In this case, quick setup will result in higher production capacity and shorter cycle times (as long as the newly available "free" time is not assigned to production). In contrast, a machine with low utilization has plenty of time for setups, and reducing setup time may have little or no economic justification. So, the financial gain from reducing setup time at a single workstation depends on how the workstation is utilized currently; how much of the available machine time is used for production and how much is used for setup? In other words, the financial benefit is inversely proportional to the time available for setups (Gallego and Moon 1992: 616).

How much time should today's factory spend on setup operations? Obviously the answer will vary depending on the particular situation, but we can offer some guidelines based on scientific principles. First, regardless of how the equipment is utilized, setup activity should never consume more than 17.5% of the total available production time (Kuik and Tielemans 1997: 178).

Consider a factory that runs two 10-hour shifts per day five days a week. During the week, five different jobs run on a particular machine, and each job requires a 4-hour setup. Therefore, the machine requires a total of 20 hours for setup opera-

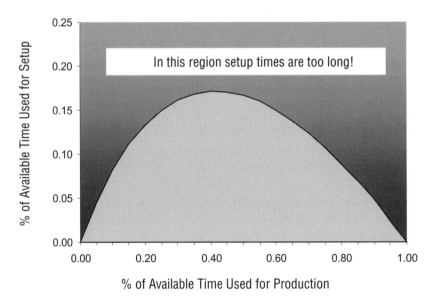

**Figure 11.1** *Upper Limit for Setup Time*

tions (5 jobs at 4 hours per setup) during a week. The factory is scheduled for 100 hours, so the proportion of available time that is allocated to setup activity for that machine is 20%. This is higher than the maximum recommended percentage (17.5%), so we conclude that the setup times should be reduced (Figure 11.1).

Now, let's assume this machine is only scheduled to run for 77 hours each week. This means the equipment is utilized for production at the very appropriate level of 77% (Chapter 12). How much time should be spent in setup activity? With a productive utilization of 77%, the setup activity should take no more than 10% of the total available time. In this case, the total available time is 100 hours, so the setup activity should take no more than 10 hours per week. Distributed over five jobs, this means setup time should be reduced to 2 hours per job.

In this chapter we show that the ultimate benefit of reducing setup time is not greater workstation efficiency but *the ability to produce in smaller lots*. We present three distinct and independent perspectives. First we review the economic order quantity (EOQ). Then we present the benefits of small lot production according to lean manufacturing philosophy. Finally we analyze small lot production from a factory physics perspective, and we discuss the financial implications.

# The Economic Order Quantity

Consider a batch production process that requires a lengthy setup operation of 3 hours. Let's assume the processing rate is 1 piece per minute. Because of the inefficiencies associated with the setup operation, management has determined that the operation should produce in lots of 10,000 pieces. Is this a reasonable policy?

| Lot Size | 100 | 1,000 | 10,000 |
|---|---|---|---|
| Setup Time | 3 hours | 3 hours | 3 hours |
| Processing Time per Piece | 1 minute | 1 minute | 1 minute |
| Effective Processing Time | 2.8 | 1.2 | 1.0 |
| Improvement | N/A | 58% | 64% |

*Table 11.1* *The Effect of Lot Size on Effective Processing Times When Setup Times are Long*

As Table 11.1 shows, when lot size is increased from 100 to 1000 pieces, the effective processing time is reduced by 58%. This is a significant improvement! By amortizing the setup time over many pieces, the effect of long setup time on effective processing time is greatly reduced. When lot size is increased to 10,000, the effect of long setup becomes almost negligible. However, the *amount* of improvement as compared to the 1000 piece lot is much smaller. In summary, *if lot sizes are already large, the benefit of increasing lot size further is insignificant.*

At the height of the scientific management movement, Ford Harris (1913) developed a mathematical model to compute the optimal manufacturing lot size. His now famous model, the *Economic Order Quantity* (EOQ), is the cornerstone of inventory management research. As with any mathematical model, EOQ is based on several simple assumptions. This is not a weakness. Harris' results are quite correct given his underlying assumptions. However, as we have pointed out (Chapter 5), using simple assumptions can restrict the applicability of any model. Specifically, Harris' assumptions include: 1) instantaneous production of an entire batch, 2) instantaneous and immediate delivery, 3) deterministic customer demand (no uncertainty), 4) constant customer demand (no variability), 5) constant setup cost, and 6) products can be analyzed independently.

To determine the lot size that yields the lowest possible production cost, Harris first developed a mathematical model for production cost.

$$Cost = \left(\frac{Q}{2}\right) \times H + \left(\frac{D}{Q}\right) \times K$$

where:  $Q$ = lot size
$H$ = holding cost
$D$ = demand
$K$ = setup costs.

We know from calculus that if we differentiate this cost equation with respect to $Q$ and set the resulting expression equal to zero, we will obtain an equation for the lot size at which cost is either a maximum or a minimum.

$$C = \left(\frac{Q}{2}\right) \times H + \left(\frac{D}{Q}\right) \times K$$

$$\frac{dC}{dQ} = \frac{H}{2} - \frac{DK}{Q^2} = 0$$

$$Q = \sqrt{\frac{2DK}{H}}$$

We leave it to the reader to verify that the second derivative of the cost equation is always positive for any positive order quantity. This means that our equation for $Q$ corresponds to the minimum production cost. Harris designated this quantity the *economic order quantity*.

$$EOQ = \sqrt{\frac{2DK}{H}}$$

To illustrate how the EOQ formula is used, we will present a simple example. Consider a factory that uses 60 doodads per week at a cost of $12 per doodad. The management estimates annual holding costs at 40% of the cost of the item being held. Setting up the machine to produce the doodads costs the factory $80.

$H$ = Holding Costs = 0.40 × $12 = $4.80
$D$ = Demand = 60 pieces per week × 52 weeks per year = 3120 pieces each year
$K$ = Setup Costs = $80

$$EOQ = \sqrt{\frac{2DK}{H}} = \sqrt{\frac{2 \times 3120 \times 80}{4.80}} = 322$$

So, the most "economical" lot size for doodads is exactly 322 pieces. We are not suggesting that our readers use Harris' formula to compute their manufacturing lot sizes. However, we can gain considerable insight about batch production by analyzing the EOQ model. Figure 11.2 is a graphical representation of the EOQ.

In this figure, the decreasing curve is the annualized setup cost. The rising straight line is the annualized holding cost, and the top curve is the total production cost according to the EOQ model. The total cost is at its minimum when the setup costs exactly balance the holding costs. This is the point where the setup cost curve crosses the holding cost line.

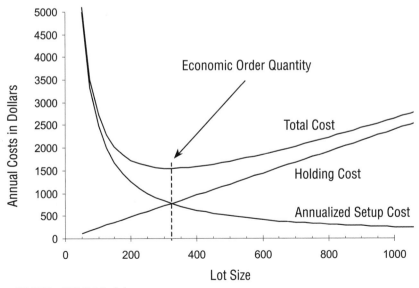

*Figure 11.2 The EOQ Model*

Now let's reconsider the example we looked at previously. If setup time is reduced to 3 minutes, as shown in Table 11.2, how much is the effective processing time improved by increasing the lot size?

| Lot Size | 100 | 1,000 | 10,000 |
|---|---|---|---|
| Setup Time | 3 minutes | 3 minutes | 3 minutes |
| Processing Time per Piece | 1 minute | 1 minute | 1 minute |
| Effective Processing Time | 1.03 | 1.003 | 1.0003 |
| Improvement | N/A | 2.6% | 2.9% |

*Table 11.2 The Effect of Lot Size on Effective Processing Times When Setup Times are Short*

In this case, effective processing time improves less than 3%, even when lot sizes increase from 100 pieces to 10,000 pieces! This phenomenon can be illustrated graphically by using the EOQ model. Figure 11.3 represents the EOQ model with three different setup costs. In summary, *if setup time is short, the benefit of increasing lot size is insignificant.*

In Figure 11.3, the optimal lot size from a cost perspective is where the setup curve crosses the holding cost line. The setup cost curves correspond to three different annualized setup costs: high setup cost, moderate setup cost, and low setup

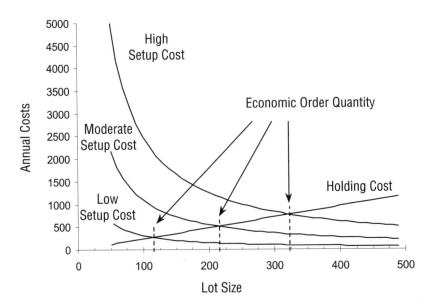

*Figure 11.3 EOQ with Decreasing Setup Costs*

cost. Clearly, *lower setup cost requires a smaller lot size for the production cost to be minimized*. Reducing setup cost without changing lot size actually causes production cost to be too high!

# Evidence from American Manufacturing

One of the best studies of manufacturing practices in America was a survey by Sakakibara, Flynn, and Schroder (1993). They identified twelve elements of JIT and determined the extent to which American manufacturers had adopted these practices. The elements they chose are listed below:

- setup time reduction
- small lot sizes
- JIT deliveries from suppliers
- supplier quality level
- small group problem solving
- training
- daily schedule adherence
- preventive maintenance
- equipment layout
- product design simplicity
- *kanban*
- pull system support.

Not surprisingly, setup time reduction correlated significantly with almost every other JIT element. In other words, manufacturing companies that practice almost any of the JIT elements also tend to practice quick setup. This implies that quick setup is accepted readily by manufacturers as a legitimate technique and is instituted early in the JIT adoption process.

It is also interesting to notice which JIT elements do *not* correlate with setup time reduction. The first JIT element that is independent of quick setup is supplier quality level. This is not surprising because setup time reduction is clearly the responsibility of the manufacturing plant, and supplier quality is achieved by working together with suppliers. The *only other JIT element that does not correlate with setup time reduction is small lot sizes*. This is very disturbing! Small lot production is the primary reason for achieving quick setup in the first place!

Our industry and consulting experience corroborates Sakakibara's findings. Almost every factory we visit has some kind of setup time reduction program under way; those that don't usually wish they did. Recently we were asked to tour a prominent automotive stamping plant to suggest ways they could reduce their setup time. We asked what management would do differently if setup time could be magically reduced by, say, 50%. We were met with blank stares, confused looks, and responses that were rooted in mass production. The tragic truth is that *most American manufacturers are trying to reduce setup time without even knowing why.*

## Quick Changeover and Small Lot Production

Figure 11.4 is a graphical representation of a manufacturing process that produces two different items (A and B) during a given time period. The items are produced in large batches, and a changeover or setup operation is required whenever production switches from A to B. Column *i* represents the initial process before it has undergone any improvement. Item A is produced in a large lot; then a lengthy changeover operation takes place. Then item B is produced in a large lot followed by another changeover operation, and the cycle repeats.

Column *ii* represents the process after setup time has been reduced. It also raises the question of what to do with the newly available time.

Column *iii* is a representation of how the time should be allocated according to conventional mass production tenets. It suggests that the time formerly used for setup operations should now be used for additional production. Consequently, this strategy results in an increased lot size for both items.

Column *iv* represents the time allocation according to lean manufacturing principles. The lot size is reduced, and items A and B are produced more frequently.

The strategies presented in columns *iii* and *iv* have individual strengths and

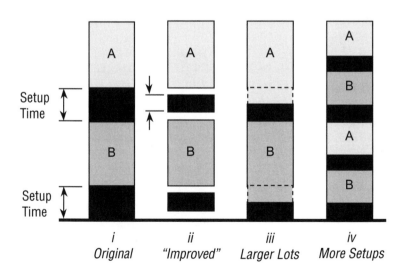

*Figure 11.4 Batch Production Options with Quick Setup*

weaknesses, but what is more important is their respective contributions to overall system performance. By examining holistically these two approaches, we will gain valuable insight into the behavior of a manufacturing system.

Again, the major justification for the strategy shown in column *iii*, the increased lot size option, is that the setup time is amortized over a larger production run, thereby reducing the effective processing time (and cost) per piece. This is a typical application of mass production philosophy, and from that perspective, it makes perfect sense. If manufacturing operations are evaluated according to their *individual efficiencies*, larger lots are often chosen because they maximize the *economies of scale*. This is an inherent strength of the large lot production strategy and the principal reason why it is in such widespread use today.

Another professed reason for choosing larger production lots is a perceived increase in quality. The logic stems from the notion that an improper setup results in poor quality. Therefore, according to large-lot proponents, once an operation is producing good parts, it should continue to produce as long as possible. Thus, the risk of producing defective parts is minimized. This perceived advantage of large lot production is a double-edged sword. If, on the other hand, the process begins to generate defects, then all subsequent parts in the lot are likely to be defective. In such cases, a large lot size implies that many defective items will be produced. This weakness of large lot production, which exposes the factory to great financial risk, is well documented in the operations research literature (Porteus 1986; Lee and Rosenblatt 1987; and Bowman 1994).

Column *iv* of Figure 11.4 represents the lean manufacturing approach: small lot production and frequent changeovers. Whereas mass production focuses on the efficiency of individual operations, lean manufacturing focuses on total system efficiency. Ultimately, the cost associated with producing an item is determined by the total system efficiency. Therefore, improving total system efficiency is an excellent way to reduce production costs and increase profits. By understanding how small lot production and frequent changeover affect total system efficiency, we can draw conclusions about how today's factory should run.

## Total System Efficiency

We know that manufacturing costs are tied closely to the performance of the overall production system. This total system efficiency is a function of several factors. One of the dominant factors is *cycle time*, which is the time the item spends in the production process. Remember that cycle time is completely different from *processing time*. Processing time is simply the length of time required to process an item at a particular workstation. Cycle time is the length of time production material spends in the system.

Again, processing time, the actual work time, only accounts for a small fraction of total cycle time. By far, the largest constituent of cycle time is waiting time. Our golf analogy illustrated this in Chapter 4. The waiting time is spent waiting in queues, waiting for tools, waiting for people, waiting for machines, waiting for material handling resources, waiting for inspection or rework, waiting for the other items in the lot to finish, and many other waiting delays.

# A Lean Manufacturing Perspective

According to lean manufacturing philosophy, there are definite advantages associated with small lot production. Most have direct impact on the total system efficiency, but some are indirect and more difficult to quantify. In this section we discuss several advantages of quick setup and small lot production from a lean manufacturing perspective.

## *Eliminates Overproduction*

Overproduction means producing more, sooner, or at a faster rate than is needed to satisfy customer demand (Chapter 7). When setup operations are long, difficult, expensive, or unpleasant, factories often produce in large lots to compensate. We are familiar with an automotive stamping plant in southeastern Michigan that produces 40,000 to 50,000 body panels in a single production run. A lot this large satisfies the requirements of an automotive assembly plant for approximately seven weeks. This means some body panels will wait nearly two months before they are used. Another plant, which supplies the home appliance industry, produces in lots as large as 2 million pieces. This is enough to satisfy customer demand for several months. An automotive components supplier in southeastern Michigan recently overproduced 200,000 components just before the product engineers changed the part design rendering the parts obsolete. If these factories had reduced setup time, they could have produced their parts more frequently and in smaller lots. In other words, their production would have *matched* more closely the customer demand. In addition, there would have been a much lower risk of damage or obsolescence. Eliminating overproduction is one of the most important reasons for reducing setup time *and* lot size.

## *Reduces Cycle Time*

Recall that cycle time is the total time material spends in the production process. Under a large lot policy, material remains in the process for a very long time. This is because each piece in the lot waits until all the other pieces are complete before it moves to the next process. This waiting time is by far the largest constituent of cycle time. Smaller lot sizes reduce unequivocally cycle time and production costs.

## Reduces Congestion

Large lots usurp production resources for an inordinate amount of time. When products are processed in large lots, other items must wait for these resources to become available. This is analogous to a long traffic light in rush-hour traffic. Drivers entering from a side street must wait for a seemingly endless line of traffic to pass before they can proceed. Smaller lot sizes reduce congestion.

## Reduces Inventory

While inventory in itself is not necessarily evil, *excess* inventory should be eliminated. Doing so can result in very significant cost savings. In fact, inventory reduction is the primary reason some companies pursue lean manufacturing. When lot sizes are reduced, there is an opportunity to reduce inventory levels. *A properly running small lot production strategy will always have less inventory than an otherwise equivalent large lot strategy.*

## Reduces Space Requirements

This advantage is actually a result of some of the others. If overproduction is eliminated and congestion and inventory are reduced, less factory space will be needed to store and handle production material. While some manufacturers argue about whether this is an actual advantage, others are realizing huge benefits. For example, Freudenberg-NOK has reduced floor space requirements by 32% through lean manufacturing improvement efforts (Liker 1997: 188). This implies the factory can increase its output substantially without additional "brick and mortar," and Joseph Day, CEO of Freudenberg-NOK, is determined to make it happen!

## Reduces Scrap

Large lot proponents argue that defective parts are the result of improper setup; therefore, once a process is producing good parts, it should run as long as possible. The converse implies that if a defect is not detected, the large lot strategy would produce a large quantity of defects. In our experience this is often the case. The automotive stamping plant mentioned above recently sent 42,000 defective trunk or "deck" lids (seven weeks' worth!) to an assembly plant. An automotive component supplier with whom we are familiar narrowly averted a similar disaster by catching in-house 100,000 defective transmission valves. Since all the valves had been made in the same lot with the same setup, all of them were defective and had to be scrapped. Obviously, these manufacturers would have been better off using small lot production.

## Improves Quality

In spite of what we would like to believe about statistical process control, most manufacturing processes do drift. This phenomenon, which causes autocorrelation in quality control data (Standard 1997), is present in more than 70% of industry situations (Alwan 1991). Porteus (1986) and Lee and Rosenblatt (1987) demonstrate that lot sizes should be reduced as the probability of the process going out of control increases (Bowman 1994). Also, with smaller lot sizes, production operators tamper less frequently with the process. This eliminates a significant source of variability and leads to better quality. (A friend of ours recently described his idea of a perfect factory as a machine, a dog, and a man. The machine would make the parts, the man would feed the dog, and the dog would bite the man if he touched the machine!)

## Improves Problem Solving

When lot sizes are large, considerable time elapses between when an item is produced and when it is used. If defects are found at the point of use, it is far too late to do anything about it. It is too late to correct the process and may be too late to repair or salvage the defective items. On the other hand, if lot sizes are small, the items are used much sooner, problems are discovered earlier, and it is more likely that the root cause of the problem will be found and resolved.

## Improves Response Time

This advantage is primarily a result of shorter cycle times. If setup times are short and lot sizes are small, the factory may be able to respond more quickly to changing customer requirements such as changes in design, material, or purchase quantity.

## Increases Flexibility

Being able to change frequently and efficiently from producing one item to producing a different item makes a factory more flexible. Manufacturing is a dynamic environment with a lot of uncertainty, and customers have a way of changing their requirements constantly. Being flexible is one way to turn unexpected changes or problems into opportunities to satisfy customers.

## Increases Production Capacity

If setup time is a substantial portion of total available production time, then capacity can be increased significantly by even modest improvements in setup time. This means the time used for production and setup is a smaller proportion of the total available time. The newly available time is a capacity "buffer" that helps the factory absorb variability and uncertainty. Higher capacity means lower capacity utilization, which translates to lower production cost (Chapter 12). If setup times are reduced enough, it may even be possible to process more jobs using the existing equipment.

### Lowers Production Costs

This topic will be treated in greater detail in Chapter 12, but it is important to recognize that cost is directly related to the *variability* in processing time (Zipkin 1995). This variability is influenced strongly by lot size, so reducing lot size directly lowers production cost. *Large lots directly increase production cost regardless of the apparent efficiency they may provide to an individual department.*

### Increases Worker Motivation

Setups are usually complicated by missing parts, uncertain procedures, heavy or awkward tooling, incessant adjustment, and very little help or understanding from management. This can be extremely frustrating for shop floor workers.

Quick setup is not about working faster or harder. It is about working *differently*. When shop floor workers and managers cooperate to find creative solutions to setup problems, each person has a stake in the improvement. This promotes loyalty and teamwork and increases motivation. When lots are large, both workers and management know that the job will not need to be set up again for a long time. This promotes apathy and procrastination about correcting setup problems since the hardship will not occur again for quite some time. On the other hand, if lots are small, then setup operations recur quite often, sometimes daily. This motivates people to solve permanently the setup problems since they know that the difficulties will otherwise occur often.

Also, the personal sense of satisfying the customer is all but lost when the factory makes thousands or millions of parts that will not be used for many weeks or months. When producing in small lots, workers feel an urgency about the current job because production is more closely linked to the customer.

## Variability, Cycle Time, and Cost

The amount of waiting time in a manufacturing operation depends primarily on three factors: *variability, utilization,* and *management policy*. Variability in this context is not related to "process variables" or quality characteristics. When analyzing production systems, "variability" refers to the variation in cycle time (Chapter 5). Consider a workstation that normally processes a part every ten minutes but can take as long as thirty minutes; the processing time varies from ten to thirty minutes. This variability is passed to the next workstation and propagates downstream. High *variability* will lead to a lengthy cycle time, a large amount of work-in-process (WIP), and severe congestion in the production flow. These effects are exacerbated if the production resources are *highly utilized* (Chapter 12).

Also, *management* can make conscious decisions that affect waiting time. For example, a manager might move a rush order ahead of another to satisfy a good customer's urgent need. These kinds of decisions can have detrimental consequences for the production flow and can cause waiting time to rise sharply. Occasionally managers *do* need to take emergency measures, but we must be cognizant of the ramifications those decisions have for the factory.

There is a direct relationship between variability and cost. An increase in processing time variability will cause production costs to rise. This phenomenon is severe in the presence of high utilization (Chapter 12). Since variability causes increases in cycle time, WIP, congestion, and cost, reducing variability is wise for any manufacturing organization attempting to sharpen its competitive edge.

## How Lot Size Influences Variability

How does lot size affect variability? We will answer this question by applying techniques from factory physics (Chapter 5) to an actual factory situation. Recall our discussion about the squared coefficient of variation, *scv* (Chapter 5). The *scv* is an objective measure of variability in processing time. We define the coefficient of variation, *cv*, as the standard deviation of processing time divided by the mean processing time. The squared coefficient of variation is simply the *cv* squared.

$$ cv = \frac{\sigma}{t} \qquad scv = \left(\frac{\sigma}{t}\right)^2 $$

An analysis of *scv* will be used to answer definitively whether a small lot policy leads to higher or lower processing time variability. This shop floor example provides valuable insight into production system performance, and builds intuition and confidence based on scientific principles.

We will consider once more a stamping facility in southeastern Michigan that produces body panels for automotive assembly plants. The panels are made on large presses, and each press makes many different kinds of panels. Stamping presses must switch from producing one kind of panel to producing a different panel fairly often. This example is directly applicable to any factory producing a variety of items in batch.

We will use the *scv* of effective processing time to compare two production policies; one is large lot production, and the other is small lot production with frequent changeovers. Using the *scv* makes the analysis valid and objective. Before we can compute the *scv* of effective processing time, we first must determine the effective processing time and the variance of effective processing time. This requires some basic information about the process.

The values shown below are typical processing parameters for a stamping operation that produces automotive body panels. However, we have simplified the numbers for clarity.

Average processing rate  = 6 hits per minute
Average processing time, $t_o$  = 10 seconds
Mild variation, $c_o$  = 0.25
Average setup time, $t_s$  = 2400 seconds (40 minutes)
Mild variation, $c_s$  = 0.25
Number of pieces in a lot, $N_s$ = 3600

Using this information, the *effective* processing time, $t_e$, can be calculated easily as shown below:

$$t_e = t_o + \frac{t_s}{N_s} = 10 + \frac{2400}{3600} = 10.7$$

So, after taking into account the setup time, the effective *average* processing time is 10.7 seconds.

Now we can use the definition of the *scv* and a little bit of algebra to compute the variance of original processing time and setup time. (The variance is simply the standard deviation squared.) The standard deviation of the original processing time is designated $\sigma_o$, and the standard deviation of setup time is designated $\sigma_s$.

$$\sigma_o^2 = c_o^2 t_o^2 = (0.25)^2(10)^2 = 6.25$$

$$\sigma_s^2 = c_s^2 t_s^2 = (0.25)^2(2400)^2 = 3.6 \times 10^5$$

There is only one more value we need before the effective *scv* can be computed: the standard deviation of the effective processing time. The standard deviation of effective processing time can be determined by the following equation:

$$\sigma_e^2 = \sigma_o^2 + \frac{\sigma_s^2}{N_s} + \frac{(N_s - 1)}{N_s^2}t_s^2$$

$$\sigma_e^2 = 6.25 + \frac{3.6 \times 10^5}{3600} + \frac{3599}{(3600)^2}(2400)^2 = 1706$$

The effective *scv* can now be computed and used as a quantitative measure of the processing time variability. The value obtained below will be used to compare the large lot production strategy with the lean manufacturing strategy depicted in column *iv* of Figure 11.4.

$$scv_e = \left(\frac{\sigma_e}{t_e}\right)^2 = \frac{1706}{(10.7)^2} = 14.9$$

So, the effective *scv* using the large lot production policy is 14.9.

Lean manufacturing principles state that if setup time is reduced, the lot size should also be reduced, consequently forcing frequent changeover operations. To maintain consistency of comparison, it is important that we reduce the setup time and lot size by a corresponding amount. We chose a 50% reduction. These are the new values for the processing parameters.

Average processing rate         = 6 hits per minute
Average processing time, $t_o$    = 10 seconds
Mild variation, $c_o$            = 0.25
Average setup time, $t_s$       = 1200 seconds (20 minutes)
Mild variation, $c_s$            = 0.25
Number of pieces in a lot, $N_s$ = 1800

$$t_e = t_o + \frac{t_s}{N_s} = 10 + \frac{1200}{1800} = 10.7$$

Notice that the effective average processing time (10.7) is the same for both strategies. This means that the long-run average processing rates for the two strategies are identical. By reducing the setup time and lot size by a corresponding amount, we actually forced this to be true.

$$\sigma_o^2 = c_o^2 t_o^2 = (0.25)^2(10)^2 = 6.25$$

$$\sigma_s^2 = c_s^2 t_s^2 = (0.25)^2(1200)^2 = 0.9 \times 10^5$$

$$\sigma_e^2 = \sigma_o^2 + \frac{\sigma_s^2}{N_s} + \frac{(N_s - 1)}{N_s^2}t_s^2$$

$$\sigma_e^2 = 6.25 + \frac{0.9 \times 10^5}{1800} + \frac{1799}{(1800)^2} (1200)^2 = 856$$

$$scv_e = \left(\frac{\sigma_e}{t_e}\right)^2 = \frac{856}{(10.7)^2} = 7.5$$

Using a small lot production policy, the effective *scv* of processing time has been reduced to 7.5.

Clearly, the effective *scv* is *much* lower for the small lot policy. This is important because it proves that *small lot production provides greatly reduced variability even though the equipment must be set up more often.* Furthermore, we know that variability in one processing station affects significantly the variability in the downstream operations (Chapter 5). This means small lot production contributes less variability to the overall system. Since small lot production reduces variability, it also causes less congestion, less WIP, shorter cycle time, and lower production cost when compared to an otherwise equivalent large lot production policy.

## Summary

Setup time reduction is one of the most popular lean manufacturing techniques. The technique itself is widely accepted in many factories today, but often the strategic intent is lost. The premise on which setup time reduction is based is frequently misunderstood or rejected. The primary reason for reducing setup time is to reduce the variability in processing time. Once setup time has been shortened, reducing the lot size is an excellent way to further reduce this variability.

It is no surprise that reducing lot size requires a reduction in setup time to maintain the same throughput. This is especially true when production resources are highly utilized. So, we can all agree that *quick setup enables small lot production.* Moreover, we have demonstrated both from an EOQ perspective and from a factory physics perspective that a reduction in setup time actually *requires* a corresponding reduction in lot size. Doing otherwise results in suboptimal factory performance due to excessive variability and cost. Therefore, *reducing setup time forces small lot production.* This may not be obvious, and it may be difficult to accept or internalize. After all, it is the antithesis of mass production and its emphasis on economies of scale. Nevertheless, the lean manufacturing bias toward small lot production is clearly substantiated by our objective analysis.

Setup is a source of consternation for many manufacturing managers and shop floor workers. Some supervisors do not like setup operations on their shifts because production output decreases. We are familiar with a manufacturing company where supervisors deliberately overproduce to avoid doing a setup on their shift. Why? If a machine is being set up for the next job, it is not producing parts. This results in lower production figures on the supervisors' daily production reports and in their annual salary review folders. Furthermore, operators themselves may prefer producing in large lots to avoid performing setups, especially if a setup is tedious, frustrating, or physically demanding.

Short setup times make today's factory more competitive, both in terms of cost and responsiveness to customer requirements. Reducing setup time is also an excellent way to increase the efficiency of a particular manufacturing process, especially if production resources are highly utilized. However, the most significant advantage of *quick setup* is that it enables *small lot production. This leads to a significant improvement in total production system efficiency.*

Running today's factory requires a holistic perspective that looks beyond individual workstations to the overall factory. The many advantages of small lot production provide compelling reasons for reducing setup time and lot size.

Factory physics, with its focus on variability reduction, provides a useful and objective way to analyze production lot sizes. The *scv* of effective processing time is an excellent quantitative measure, and it can be used to evaluate objectively manufacturing policies. The *scv* of effective processing time can be computed easily using the simple algebraic equations presented in this chapter. This chapter provides empirical and rigorous analytical proof that small lot production with frequent setups increases the flexibility and profitability of today's factory!

# Chapter 12
# Measuring Today's Factory

*"All things flourish where you turn your eyes."—Alexander Pope*

## Introduction

In Chapter 6 we emphasize that manufacturing strategy should be aligned with the overall corporate strategy. This ensures that manufacturing decisions support strategic business goals. Oddly, the goals and measures that manufacturing organizations establish for themselves are often *not* aligned with sound business strategies. For example, the board of directors might be concerned about the strength of the next monthly financial statement or reaching a certain share price by year's end. A manufacturing manager might be expected to improve return on investment (ROI) or return on net assets (RONA) in the following quarter. These measures may be appropriate for monitoring long-term success, and it is certainly important for today's factory to track its fiscal performance; however, as short-term measures, they often lead to a myopic manufacturing management style (Nahmias 1997: 2). "Watching the bottom and top lines, stock prices, cost per unit, and other such measures are valid as infrequent indicators of the health of entire systems—not for making tactical decisions or judging monthly or quarterly performance" (Schonberger 1996: 17).

Performance measures are much more than a way to monitor work efficiency and departmental output. They actually influence the way the factory operates and the way people do their jobs. Performance measures affect what kind of culture the factory develops, what attitudes people have, what approach is used to solve problems, what behaviors are encouraged and rewarded, and what decisions are made from the shop floor to the boardroom. Essentially, "what you measure is what you get." Therefore, great care should be taken when establishing performance measures to ensure that they support the overall business goals.

So, performance measures have an even broader purpose than keeping track of what takes place in our factories. They serve three very important functions:

1. Motivate desired behavior
2. Support management decision making
3. Assess and evaluate the status of personnel, departments, and the overall factory.

Factory performance measures often focus on *segments* of the system (for example, the efficiency of a welding operation) at the expense of *overall* efficiency and productivity. For example, managers are frequently evaluated on the efficiency of their individual departments rather than on their contributions to total factory performance. This can easily and understandably lead a manager to act in self-interest and do what is best for his department and for his career rather than what would be best for the company.

We are acquainted with a very successful manufacturing manager at an engine plant in southeastern Michigan. His department must produce 1800 engines per day to meet the performance measures on which he is judged. The assembly plant that uses these engines can use no more than 1600 engines per day. So, what does the manager do? Even though he *knows* it would be more profitable for the company to produce only the number of engines required, his plant produces at least 1800 engines every day. This is an example of good old-fashioned overproduction (Chapter 7). The extra engines accumulate by the thousands until finally a decision is made to close down the engine plant for a week or two. The performance measurement system forces this manager to make decisions that he knows are not in the best interest of the company.

The absence of a performance measure can also influence manufacturing decisions. For example, we know the manager of one of America's top stamping plants. This manager has been working to reduce the time required to set up his presses. He is doing this to achieve improved flexibility. The inventory reduction that can accompany quick setup (Chapter 11) is irrelevant to him because his "performance" is not evaluated based on the amount of inventory in the system. His performance is measured by other factors, primarily his ability to keep assembly plants supplied with a variety of body panels.

Another example involves the purchasing department of the same company. Purchasing personnel are rewarded based on reducing the price of procured items. This may seem like a reasonable goal; after all, paying less for purchased items should lead to higher profit, right? However, in this company, transportation costs are charged to the assembly plant, so they are not a consideration when the purchasing department selects suppliers. This leads purchasing to make decisions that might save a penny per unit on the purchased price while incurring a five-cent increase in unit transportation cost for the assembly plant. The purchasing department receives accolades for cutting costs while the overall production cost of the automobile rises and the profitability of the company suffers.

In this chapter we consider some traditional performance measures and evaluate their appropriateness for today's factory. Capacity utilization, one of the most popular performance measures, is examined from a factory physics perspective, and we discuss a *capacity utilization* paradox. We develop some guidelines for measuring performance, and we offer several "enlightened" performance measures that are particularly relevant for today's factory.

## Traditional Performance Measures

There are many traditional measures that manufacturing companies use to assess the performance of their factories. These include holistic, corporate measures that assess entire divisions as well as individual measures that assess the diligence of specific production workers. Every manufacturing company has its own preferred or favorite performance measures, and many companies invent their own measures to capture some important and unique aspect of their operation.

It is important to keep in mind the very critical role performance measures play beyond assessment. Performance measures are only beneficial long term if they encourage the right behavior and assist management in making the right decisions. The appropriateness of any performance measure can be evaluated by whether or not it leads to supportive behavior and sound management decisions. We have chosen three common performance measures for consideration: standard labor variance, overhead absorption, and return on investment. For each one, we demonstrate how a *critical thinking* approach can help determine whether a particular performance measure is a useful tool for today's factory manager.

### Standard Labor Variance

Scientific observation and measurement of work are not new. For years engineers have tried to determine standard and optimal ways for people to do their work. Leonardo da Vinci studied the art of shoveling more than 450 years ago (Hicks 1994: 6). Early in the 20th century at Bethlehem Steel, Fredrick Taylor, the "father" of industrial engineering, also analyzed shoveling rates. (Why all the fuss about shoveling? Well, we can't speak for Leonardo, but for Taylor, shoveling was the most prevalent work in a steel mill in those days.) Taylor found that the same size shovel was being used regardless of the weight of the material being shoveled. By determining the optimal weight of a "shovelful" and by designing appropriately sized shovels for different jobs, Taylor was able to reduce the number of men shoveling from 500 to 140. Taylor relentlessly pursued the best way for people to do their work. He honed manual tasks to their highest possible efficiency by scrutinizing each element of work and eliminating all wasted effort. The work standards he developed represent "the work rate that should be attainable by a first-class man" (Hopp and Spearman 1996: 30).

Factory managers have been following the trail blazed by Taylor ever since. We compute the standard rate at which workers should be doing their work and assess performance against that standard. We also use this technique to evaluate the output of entire departments, and we hold the area manager responsible for any shortcomings. How do we know whether the amount of work is up to standard? We simply compare the amount of work done to the amount that should have been done given the standard production rate and the number of hours worked. This comparison is known as *standard labor variance*, and it is probably the most widely used performance measure in factories today.

Is standard labor variance an appropriate measure for today's factory? Before answering this question, we define an *appropriate performance measure* as one having these three characteristics.

1. The performance measure must be *objective, precisely defined, and quantifiable.*
2. The performance measure must measure something that is *within the control of the people or department being measured.*
3. The performance measure must *encourage appropriate behavior*, in other words, behavior that supports overall efficiency and the strategic goals of the company.

Certainly standard labor variance satisfies the first criterion: it is measurable. It is usually expressed as a percentage and is defined as the actual production output divided by the expected production output based on the standard production rates and the number of hours worked, minus 1.0. Many factories generate a daily or weekly report of the standard labor variance for jobs, product lines, departments, or individual workers.

Whether or not standard labor variance meets the second criterion is not so clear. Obviously workers can directly influence efficiency and productivity; however, the ability to meet the production standard is often *not* within the control of the shop floor worker. Faulty production equipment, poor quality materials, and late deliveries are more common than recalcitrance or lack of worker diligence. An estimated 85% of production variation is caused by system faults, which are the responsibility of management, not operators (Inman 1993: 43). In such instances, blaming the shop floor workers for missing their standard output is like blaming the ballplayers for an evening shower that rains out their ballgame.

On the other hand, if the cause of the variance is recalcitrance, the supervisor already knows this, and the variance report is "unnecessarily and redundantly superfluous!" Reports generated a day or a week later will not provide timely and helpful feedback to the supervisor nor will they win back the lost production.

> Financial accounting measures lag performance because they are historical in nature, by definition reporting on activities that have already occurred. For this reason, they are irrelevant in guiding managers in their quest to improve current and future operations (Clinton and Hsu 1997: 18).

Trying to monitor worker diligence with standard labor variance reports does not meet the second criterion very well. In fact, if the variance is caused by something that is out of the control of the shop floor worker or supervisor, then the performance measure violates our second criterion completely.

Does standard labor variance encourage appropriate behavior? Let's take a look. First, standard production rates are usually determined by engineers in an office or laboratory. They are often calculated from standard time and motion tables, and they may not reflect accurately the actual production process. It is highly unlikely that the standard production rate matches the true production rate. It is even more unlikely that the standard production rate matches the customer demand rate. If the standard production rate is higher than customer demand, this performance measure leads to overproduction (which we have defined as producing more, sooner, or faster than required by the customer or next process). If the standard is lower than customer demand, workers have a tendency to *ignore* the customer and to work at the standard rate.

When standard labor variance is used to evaluate departmental performance, managers may produce in larger lots, avoiding setup operations that would usurp "valuable" production time. (As we saw in Chapter 11, large lots are responsible for considerable variability, congestion, and long cycle times, and they should be avoided wherever possible.)

Using standard rate as a performance measure encourages managers to produce the "easier" jobs early in the month, ignoring other jobs that might have higher priority. This behavior leads to excellent measured productivity (standard labor variance) throughout *most* of the month. However, it definitely distances the production process from the customer since the production schedule is based on ease of manufacture rather than on customer demand. So, an effect of this performance measure is that the shop floor workers and managers become decoupled from the customer.

Using standard labor variance also encourages postponing maintenance activities so the equipment can be utilized productively as many hours as possible before incurring downtime. In the long run, this leads to severe equipment problems, excessive downtime, and extremely high production variability. Focusing on standard labor variance can also encourage hiding defects and using inferior or defective material since quantity rather than quality is emphasized.

Principles we emphasize in this book are variability reduction, a strong customer focus, making only what is needed when it is needed, ensuring quality at the source, and maintaining the proper operating condition of production equipment. The behaviors encouraged by standard labor variance are the antithesis of these principles. We conclude that *standard labor variance is not an appropriate measure for today's factory*. If it is not appropriate, why is this performance measure so widely used? Mainly because most factories absorb their overhead expenses through labor.

## *Overhead Absorption*

Using cost accounting methods, the cost of making individual products can be calculated. This information is used to make business decisions such as how much to charge for a particular item, which items are unprofitable, which items should be discontinued, which production lines should be targeted for improvement, and whether some components should be out-sourced. In theory, determining the cost of producing an item is simple. In practice, it is inexact, difficult, and frustrating.

Direct costs are by far the easiest to determine; the precise cost of raw material, components, and purchased subassemblies is assigned to specific products. Direct labor is more complicated. A shop floor worker often works on several different products during a single shift, and the time spent working on each product has to be recorded. The labor cost is then calculated and apportioned to the items produced by the worker during the shift.

A worker who works on many different products may spend as long as 30 or 40 minutes at the close of every shift filling out a time sheet. This is a significant amount of time! Time spent recording labor data is time that is not spent producing goods, so this activity actually diminishes the effectiveness and productivity of a worker. (At one factory, we estimated that filling out time sheets consumes an entire week's labor per year.) Nevertheless, the procedure is relatively straightforward, and many manufacturing companies use this type of system to allocate direct labor.

The real difficulty lies in allocating overhead for the factory. Overhead, also known as burden or fixed cost, is the cost that does not vary with the production of different items. The salary of executives and staff, the investment in research and development, carpeting the cafeteria, paving the parking lot, and the mortgage on the property are all examples of fixed cost. These costs will be incurred regardless of what products the factory is making. They are simply part of the cost of doing business. Overhead can be considered an *indirect* cost of making a product. The challenge is to determine how these indirect costs should be apportioned to the various items produced by the factory.

The traditional approach is to allocate the overhead costs in the same proportion as the labor hours spent making each product. This approach makes sense, especially if the cost of labor represents a substantial percentage of the total cost of producing an item. This was the case a century ago when these "modern" accounting methods were developed. At that time, labor accounted for approximately 90% of the total production cost (Johnson and Kaplan 1987).

In today's factory direct labor is still a significant proportion of overall costs in some industries, but it is becoming less important in others. Hopp and Spearman have observed that "today, direct labor constitutes less than 15 percent of the cost of most products, and hence the traditional methods have been increasingly challenged as inappropriate" (1996: 205). Some companies have realized this and are introducing new accounting methods that are better suited to their manufacturing situations.

The following example illustrates the fallacy of allocating overhead costs based on labor. Consider a factory that produces two items.

1. The first item is a newly designed product with technologically advanced features and a highly automated manufacturing process. The newness and technological features of the product require the involvement of product engineering, quality assurance, and purchasing departments. The automated manufacturing process provides very rapid production rates, but equipment problems persist, so several talented manufacturing engineers are involved as well.

2. The second item is an older design that has been in production for several years. Over time, the chronic production problems have been solved, and now the item is manufactured with little involvement from engineering. The newer product was expected to replace this older design in the market, so the company did not automate the production of this older item. The processes are performed largely by hand, and the production rate is slow compared to that of the new product.

In this example, allocating fixed cost on the basis of labor would not make any sense. The first item requires very little labor since the process is highly automated; yet, it uses valuable resources of the factory (engineering, quality, and purchasing departments). The second item uses much more labor; yet, its demand on other company resources is extremely limited. Allocating the fixed cost based on labor hours is "charging" the older product for "services" it did not receive.

Another way to allocate costs is known as *activity based costing* (ABC). Using this approach, cost is allocated based on the resources that a product actually consumes or the activity required by a particular product. ABC is more accurate about "charging" items for services they *did* receive, but it is not without its difficulties. Womack and Jones acknowledge that ABC is a "great advance," but they also imply that it is not enough (1996: 262). Hopp and Spearman remind us that "it is by no means a panacea" (1996: 205). Even sophisticated cost models can be inaccurate and misleading!

### Return on Investment (ROI)

ROI is defined as profit divided by investment.

$$ROI = \frac{profit}{investment}$$

There are two ways to improve ROI. One way is to increase profit by improving efficiency, productivity, quality, or market share and by not wasting production resources. Improvements such as these are difficult to achieve quickly and tend to be long term. The other way to improve ROI is to reduce expenditures by cutting investment in employees, research, facilities, equipment, technology, or other assets. This course of action is far easier and has faster results, but it can be disastrous to a manufacturing organization in the long run.

Return on investment (ROI) is not a new measure. Pierre Dupont (1870–1954) pioneered its use in the early twentieth century. "Perhaps Dupont's most influential innovation . . . was the refined usage of return on investment (ROI) to evaluate the relative performance of departments" (Hopp and Spearman 1996: 37). ROI is a measure that is characteristic of mass production philosophy and the associated economies of scale.

Let's consider a hypothetical situation: a dedicated machine that runs as fast as possible every available minute. It never stops, even for preventive maintenance or changeover operations, as it continues to produce items that will be stored in inventory. Since inventory is viewed as an asset (or revenue) in most accounting systems, this machine could achieve a tremendous ROI. This production policy typifies the mass production paradigm.

ROI is a useful measure of overall fiscal performance, but it is extremely limited as a management tool. Focusing on short-term returns cannot lead to long-term competitiveness. A machine that produces in small lots, changes frequently the items produced, and stops regularly for preventive maintenance is much more appropriate and profitable for today's factory. Such a machine would not necessarily generate a very high ROI, but it would provide a tremendous competitive advantage for the factory!

## Capacity Utilization Paradox

Consider a company that uses return on investment as a primary performance measure. One way to improve ROI is to maximize the revenue generated by the assets by utilizing production resources to the fullest extent possible. This results in *high capacity utilization*, which, in turn, can bring about a high return on investment. Because of this, many manufacturing companies use capacity utilization as a performance measure. Moreover, some of these companies state explicitly that the pursuit of 100% capacity utilization is a primary management goal. The 24-hour factory that runs 7 days a week is often heralded as a mark of excellence. Is the pursuit of high capacity utilization a good strategy for today's factory? Let's take an objective look.

Recall the three criteria established earlier in this chapter for evaluating a performance measure. 1) The performance measure must be objective, precisely defined, and quantifiable. 2) The performance measure must measure something that is within the control of the people or the department being measured. 3) The performance measure must encourage appropriate behavior. These criteria can be used to assess whether or not high capacity utilization is an appropriate performance measure for today's factory.

First, capacity utilization is "measurable." It can be computed easily for individual pieces of equipment, production departments, or entire companies. Second, capacity utilization is clearly under the control of management. Management can establish maintenance and production schedules to achieve almost any desired level of capacity utilization. So, from the perspective of the first two criteria, capacity utilization certainly qualifies as a reasonable performance measure for the factory. The third criterion states that an appropriate performance measure must encourage behavior that supports overall efficiency and the strategic goals of the company. Whether or not the third criterion is met is not immediately clear because capacity utilization affects overall factory efficiency in ways that may not be obvious.

On the surface, high capacity utilization *seems* to be the epitome of efficiency. After all, what could be more efficient than having production resources such as workers and equipment busy 100% of the time? Because the subtleties of capacity utilization are not immediately obvious, we will use factory physics to determine objectively whether it is or is not an appropriate performance measure for the factory.

Lean manufacturing and factory physics emphasize the importance of shortening cycle time, which we have defined as the amount of time that material spends in the production process. Cycle time can be used to evaluate alternative manufacturing policies such as which machine to buy, what kind of maintenance schedule to follow, and how fully our production resources should be utilized. *A policy that leads to shorter cycle time is usually very appropriate for today's factory; one that leads to longer cycle time is not.* So, in order to understand how utilization affects factory performance, we must first understand how utilization affects cycle time.

If we regard a production operation as a series of individual workstations, each with its respective queue, then an entire factory can be visualized as a queuing network and analyzed accordingly. We can easily model such a queuing system using a well-known approximation to the G/G/1 queue (Kingman 1961: 902-4). This model is quite general and places no undue restrictions on the distributions of inter-arrival times and processing times. It also happens to be exact when the interarrival times are distributed exponentially (Hopp and Spearman 1996: 277). Using this model, we can compute the amount of time material spends in the system. While this may seem somewhat abstract, it is scientifically valid, and it is quite useful for analyzing and understanding the behavior of manufacturing systems.

Equation 12.1: $\quad CT_q = \left(\dfrac{c_a{}^2 + c_e{}^2}{2}\right)\left(\dfrac{u}{1-u}\right)t_e$

where: $CT_q$ = the cycle time in queue for a single station
$\quad\quad c_a$ = the coefficient of variation for arrivals
$\quad\quad c_e$ = the coefficient of variation of the effective processing time
$\quad\quad u$ = utilization (arrival rate divided by processing rate)
$\quad\quad t_e$ = the effective processing time.

Equation 12.1 gives the cycle time in queue for a single station. To determine the cycle time for multiple processes, we need a linking equation as well.

Equation 12.2: $\quad c_d{}^2 = u^2 c_e{}^2 + (1 - u^2)c_a{}^2$

where: $c_d$ = the coefficient of variation for departures
$\quad\quad c_e$ = the coefficient of variation of the effective processing time
$\quad\quad u$ = utilization (arrival rate divided by processing rate)
$\quad\quad c_a$ = the coefficient of variation for arrivals.

Equation 12.2 gives the coefficient of variation for the departure from a single processing station (Chapter 5). If we realize that this departure variability is equivalent to the arrival variability at the next process, we can use these equations to compute the total cycle time for an entire production routing.

The following example provides *extremely important insight into how variability, cycle time, and capacity utilization are related.* Consider a manufacturing operation comprising four distinct processing stations. We can ignore the time spent transporting the workpiece from station to station without any loss of generality. Now, assume each station requires an average of 1 minute to process a workpiece. In a deterministic world, it would take exactly 4 minutes for an item to pass through all four stations.

However, we know that today's factory exists in a highly variable or *stochastic* world. This means that simple addition will not provide an accurate estimate of the true cycle time. Instead, we must use a model (such as the approximation given in Equation 12.1) to determine the actual cycle time for the system. Using this approach, we compute cycle times for three levels of variability and a wide range of utilization levels. The results are presented graphically in Figure 12.1.

At low levels of capacity utilization, the system behaves very much like a deterministic system. The cycle time is only slightly above the lowest possible time of 4

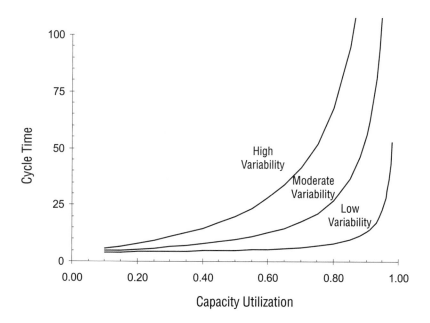

*Figure 12.1* The Behavior of Cycle Time Under the Influence of Variability

minutes. However, as utilization increases, cycle time also increases, and at high utilization levels, the cycle time increases rather dramatically. The effect is especially significant above 80%. This means *the amount of time required for material to flow through the system is much longer than normal if the workstations are highly utilized.* By Little's Law we know that long cycle times correspond to high WIP levels, so high utilization causes an excessive buildup of WIP. Yet, many factories continue to focus on high capacity utilization as a management directive. This is terribly disturbing!

Figure 12.1 also illustrates how variability in the production process affects cycle time. Recall that in this context variability does not refer to dimensional aberrations but rather to the variation in processing time, which is inherent to some degree in any manufacturing process. This variation can take the form of equipment downtime, large production lots, excessive setup time, uneven production rates, batch material movement, or any other occurrence that disturbs the smooth and steady flow of material through the system. High variability exacerbates the effects of high utilization and causes cycle time to increase dramatically. Multiple-station lines are more susceptible because the variation propagates throughout the production stream. This implies that improvement efforts focused on stations that are earlier in a routing will have a more significant effect on cycle times than an equivalent improvement later in the same routing.

In Figure 12.1 the curve that corresponds to *high variability* begins rising significantly at 40% utilization. At 60% utilization the cycle time has more than doubled from 4 to 10 minutes. When utilization is 80%, the cycle time is more than five times higher than its original value. At 90% utilization the cycle time has risen by more than a factor of ten! The *moderate variability* curve is a little slower getting started, but, once the utilization level reaches 80%, the cycle time rises quite dramatically as well. The *low variability* curve is shown as an *ideal* or *best practical case.* Most factories do not operate anywhere near the low variability curve. So it is plain to see that at utilization levels above 80%, cycle time becomes unacceptably long. As expected, each of these curves, even the low variability curve, tends toward infinity as utilization approaches 100%. As a point of reference, many company policies dictate a capacity utilization level well in excess of 80%, and some target 100% utilization as an explicit company goal.

From the analysis presented above, a paradoxical conclusion emerges: *high capacity utilization is not a justifiable goal for today's factory!* This well-supported position is widely held in the operations research community. For example, Hopp, Spearman, and Duenyas make the following observation: "The desire for greater short term profits causes many manufacturing operations to load to full capacity even though it would be more profitable to run at lower production rates" (1993: 71). Lowering the utilization of production resources can actually improve the overall efficiency and profitability of a manufacturing operation.

To understand *conceptually* why utilization affects cycle time so dramatically, consider the following example. Figure 12.2a shows a production line consisting of four similar workstations. Each workstation is lightly utilized and consequently has plenty of excess capacity. Even though processing time varies randomly at each workstation, the excess capacity can compensate easily for any disruptions to the production flow.

Now consider a production line that is managed differently. The workstations shown in Figure 12.2b are highly utilized and therefore have very little excess capacity. A small amount of excess capacity cannot possibly compensate for even minor variations in processing time. Therefore, the variation causes WIP to accumulate between the workstations. This increases the amount of time it takes for material to flow through the line because one particular workpiece cannot be processed until all previous pieces have been processed. Also, variability in the first station propagates through the line causing more severe variation in subsequent workstations. This illustrates how variability *and* utilization can affect cycle time, and provides some intuitive understanding of Figure 12.1.

Direction of Flow

*Figure 12.2a* Work Flow in a Line With Low Capacity Utilization

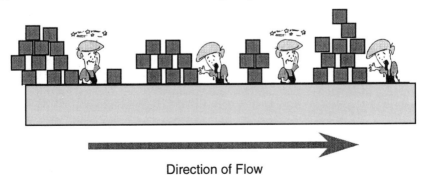

Direction of Flow

*Figure 12.2b* Work Flow in a Line with High Capacity Utilization

Clearly, high utilization leads to dramatic increases in WIP and cycle time. We have demonstrated this through rigorous analysis and conceptual reasoning. Long cycle time is a proven cause of severe problems such as poor quality, congestion, little flexibility, and long lead times. At this point, it might seem reasonable to conclude that high utilization is highly overrated.

What about profitability? We still have not demonstrated *how utilization affects profitability*. The relationship between utilization and profitability is of the greatest importance since most factories exist to make money. To understand this relationship, the production system must be analyzed from a cost perspective. Fortunately, a reasonably simple way to do this has been developed (Zipkin 1995). Zipkin developed a performance index that can predict the behavior of a production system over a wide range of situations. This performance index (which he denotes as $\sigma$) can be used as a surrogate for the actual cost performance of a production system. Zipkin shows that optimal total cost is the product of $\sigma$ and certain other terms. These other terms depend only on cost factors, not on system parameters such as production rate and customer demand, so system performance is essentially proportional to $\sigma$.

Any production policy that causes $\sigma$ to rise also causes the production cost to rise. Higher cost translates to lower profit, so policies that cause $\sigma$ to rise should be avoided. Conversely, a production policy that decreases the value of $\sigma$ yields higher profit through decreased production cost. The production system factors that cause the value of $\sigma$ to rise are high capacity utilization and high variability in processing time. Specifically, utilization (defined as the demand rate divided by the processing rate) and the *scv* of processing times are the primary factors that influence $\sigma$. So we see that profitability is increased when a production system has low capacity utilization and low processing time variability. The same two factors that can be used to reduce WIP and cycle time can also be used to increase profitability!

*So we conclude quite definitively that capacity utilization is not an appropriate performance measure for today's factory since profitability degrades significantly when either utilization or variability is too high.* As a point of interest, this analysis considers only "hard" or verifiable costs such as inventory holding costs and back-order penalty costs. Therefore, our conclusions are very conservative and do not take into account the many other "soft" benefits associated with reducing cycle time.

## Enlightened Measures

We have examined four traditional measures: standard labor variance, overhead absorption, return on investment, and capacity utilization. At the beginning of this chapter, we stated several very important functions that performance measures serve: motivate desired behavior; support management decision making; and assess and evaluate personnel, departments, and the factory as a whole operation. According to these criteria, none of the traditional measures that we have analyzed is a relevant, usable management tool. This is primarily because these measures do not encourage actions and decisions that lead to success for the factory (Clinton and Hsu 1997).

How should today's factory measure its performance? There isn't a generic answer to this question (Spoor 1998). Instead, we suggest that a company develop its own performance measures based on its own overall objectives.

Each company must develop its own metrics based on its own individual strategy. Moreover, the metrics used should be well thought out to ensure that they can be linked to strategic objectives and linked together in such a way that they reinforce each other (Clinton and Hsu 1997: 22).

We have compiled a short list of performance measures that companies are using to manage their factories. Some of these have wide applicability and can be adopted by almost any company wishing to encourage improvement. Others are more case specific, and are offered to help readers generate their own ideas for measures that are relevant to their specific circumstances.

1. Customer lead time (order to delivery)
2. Cycle time (throughput time, sojourn time, or "dock-to-dock" time)
3. Inventory (measured in time units)
4. Inventory turns
5. WIP turns
6. Customer service (delivery performance)
7. Number of stock-outs (as a percentage of orders shipped)
8. Shipments per day
9. Customer satisfaction (number of customer complaints)
10. Market share
11. Percent of revenue from innovation
12. Number of new products introduced
13. Supplier lead time (order to delivery)
14. Supplier delivery performance (percent on-time delivery)
15. Supplier quality (percent defects from suppliers)
16. Average transaction cost per supplier
17. Average transaction cost per delivery
18. Number of suppliers
19. Number of local suppliers
20. Number of purchase orders

21. Deliveries per day
22. Supplier satisfaction (number of supplier complaints)
23. Quality (ppm to customer)
24. Quality (in-house scrap rate)
25. First time yield rate (percent of items produced without rework)
26. Appraisal cost per defect found
27. Profitability
28. Profit per square foot
29. Profit per employee hour
30. Equipment availability (NOT equipment utilization)
31. Warehouse space
32. Distance traveled
33. Setup time (should be coupled with lot size)
34. Lot size
35. Production output (should be coupled with customer demand)
36. Productivity (pieces per employee hour)
37. Employee satisfaction (number of employee complaints)
38. Employee participation (percentage of employees writing suggestions)
39. Absenteeism (percent of workforce absent)
40. Safety (number of days between lost-time accidents)
41. Number of in-house job classifications
42. Average number of jobs workers are trained to do
43. Number of actions requiring management approval
44. Percent of value-added activity
45. Percent decrease in nonvalue-added activity

## Summary

The resurgence of management policies based on short-term measures such as return-on-investment is a disturbing development. Focusing on such short-term goals and having a limited outlook can render a manufacturing organization unresponsive and unprofitable. For example, to maximize ROI, a manager might choose to overutilize production resources, thereby maximizing revenue from the asset. This leads to lower profit and a proliferation of other detrimental effects. While many managers are aware of the incongruities in their performance measures, some find it difficult to accept that individual or departmental "efficiency" does not necessarily improve corporate performance. We emphasize that optimizing individual segments of a production system does not lead to an optimized whole.

The lean manufacturing paradigm is a useful framework in which to view management issues such as capacity utilization. In Chapter 4 we defined the fundamental goal of lean manufacturing: "*to achieve the shortest possible cycle time by streamlining the flow of production material throughout the value stream.*" Cycle time is affected dramatically by capacity utilization, and variability exacerbates the adverse effects of high utilization (Kannan 1998: 578). This variability can take the form of batch production, batch conveyance, uneven production rates, equipment downtime, excessive setup time, poor quality, or any other occurrence that disturbs the smooth and steady flow of material through the production process. This leads to two important guidelines: 1) reduce variability wherever possible throughout the production process; 2) do not strive for 100% capacity utilization. Utilizing resources at levels greater than 80% of capacity causes dramatic increases in WIP, cycle time, and cost.

While these guidelines are a natural result of applying lean thinking to the issue of capacity utilization, they can also be discovered through factory physics and a rigorous analysis of production costs. The optimal average cost in a production system is a function of capacity utilization and variability in the production process. An increase in either of these parameters causes WIP, cycle time, and production cost to rise. The lean strategy of reducing cycle time also results in higher profit for today's factory. This provides compelling insight into lean thinking and why it is so powerful.

*Chapter 13*

# Transforming Today's Factory

## Introduction

Can the manufacturing *principles* and *practices* presented in this book be trans-ferred to an *existing manufacturing enterprise?* The answer is a resounding yes! These concepts and methods are transferable to virtually any size company in any industry in any cultural setting. That is not to say transforming today's factory is necessarily easy, only that it is possible. The transformation process requires hard work and careful planning. The benefits, however, are well worth the effort.

An international study of work organization, production control, parts procure-ment, and labor relations concluded that indeed *these methods are highly transfer-able*, although they must be adapted to fit the management environment in each location (Kumon 1994). We agree that the "methods" are transferable, but we add this *caveat*: it is the principles, not necessarily the practices, that apply universally. Of course, adapting *management style* to fit the cultural environment is perfectly appropriate (Smith 1998).

While it is dangerous to apply the practices indiscriminately, it is even more dangerous to modify the principles to fit an existing corporate culture. Doing so simply results in a new name for the same anachronistic production strategies that led to noncompetitive performance in the first place. It is illogical to do the same things under a new name and expect different results.

We have found that transforming today's factory is a journey, not a destination. It is important for management to have both long-term and short-term goals and milestones along the way; however, the transformation process is never-ending. This chapter ties together many concepts presented in previous chapters and provides a guide for the factory transformation process.

# Getting Started

Once the corporate goals are defined and the manufacturing strategy has been established, the factory is ready to begin transforming itself. This raises two questions that must be answered: 1) where to begin and 2) how to get started. After all, it is important to have the first efforts produce visible and meaningful improvement.

Where should the factory's transformation begin? We strongly suggest that a product family be chosen for initial transformation. By "product family" we mean a group of similar products that flow, more or less, through the same set of processes as raw material is transformed into a finished product. There is no reason to shy away from a high-volume product family or even the factory's main product line. Starting with a high-profile product family highlights the importance of the transformation and leads to meaningful and highly visible improvements.

Note that we do not suggest starting in a specific area of the factory. A small-area approach is appealing because it seems contained and manageable. However, there is no reason to limit the improvements to a single area. Even if the transformation is a stunning success, the factory has merely created an island of excellence. There has been no improvement in the overall flow of production material, and the factory will not benefit significantly from isolated improvement efforts. Remember that optimizing pieces of the system does not add up to an optimized whole (Taylor 1996; Hobbs 1994: 44).

How should the factory's transformation get started? It is difficult to discuss a factory's transformation in a generic way. We feel strongly that superior performance cannot be achieved by using a "blueprint" of another company's manufacturing system. "It should be clear by now that the solutions of the future cannot be 'bought off the shelf,' but must be pioneered" (Warnecke and Huser 1995: 43). Nevertheless, a *framework* for the transformation process may help our readers visualize how their own factory transformations might progress.

*A Transformation Framework:*
1. Understand the business or corporate goals.
2. Develop a manufacturing strategy that supports the corporate goals.
3. Choose a change agent.
4. Establish a guiding coalition.
5. Assess the current state of the factory.
6. Develop a near-term and long-term vision of the future state.

7. Communicate that vision to every employee in the company.
8. Develop a transformation plan.
9. Manage the transition.
10. Make it a way of life for all employees.

## Managing the Transformation

Today's factory will not be transformed by magic. There is no potion to drink or incantation to recite. A transformation involves hard work, sound strategy, the commitment of management, and an appropriate organizational structure. Above all, *the transition must be managed effectively.* Fortunately, there *are* effective ways to manage the transformation process. William Bridges has described very clearly, accurately, and sensitively the transition process and has presented thoughtful and practical suggestions for making it a positive time of change. We use his methods extensively in our consulting work, and we draw from his book, *Managing Transitions,* in this section. The factory transformation is characterized by three distinct phases: the *ending,* the *neutral zone,* and the *new beginning* (Bridges 1991).

Contrary to popular belief, the starting point of a factory transformation is not the development of a *future-state vision* or the desired outcome that is projected onto the factory. The starting point of the transformation is the *ending* that must come if the current state is to be left behind. The demarcation of the "end of the old" acknowledges what most people are feeling—the end of their jobs as they were. Without an ending there cannot be a new beginning (Bridges 1991: 19-30).

After letting go of the old situation, there is a transitional period before the new order is truly established. This is known as the neutral zone. The neutral zone is a time of great uncertainty and even fear. The old methods are gone, and workers and managers are still uncomfortable with the new methods. In the neutral zone, people miss what they enjoyed and liked, and they don't know exactly what to expect of the new jobs. It can be an emotional time. Management can offer reassurance by formally recognizing the difficulties that all employees are experiencing. Interestingly, the neutral zone is a creative period when major innovations are most likely, and revitalization of the organization begins (Bridges 1991: 36-7).

After some time has been spent in the neutral zone, the company can anticipate a new beginning. It may seem counterintuitive to begin the transition with an ending, and to end the transition with a beginning, but it is the single most effective way to manage the transformation process. Understandably, most companies attempt to start the transformation with the new beginning. This leads to resistance, serious morale problems, and little if any measurable improvement. The transition can be considered complete only after the new beginning has been reinforced and has become a way of life (Bridges 1991).

## *The Ending*

Even before letting go of the old situation, there are two important steps management should take: 1) determine specifically how people's behavior and attitudes will need to change in the new scheme of things, and 2) *sell the problem* to everyone in the organization. The first step is essential for understanding any cultural issues that might need to be addressed. The second step ensures a common understanding of *why* the transformation must take place.

It is very difficult for human beings to accept solutions to problems they do not see. This quirk of human nature is easily observed by noticing how many people are caught in summer rainstorms without umbrellas. On a sunny day it is difficult for most people to take an umbrella with them; the umbrella is a solution for a problem that cannot be seen.

Similarly, transforming the factory involves solving problems that have not previously been seen as obstacles. For example, having an overabundance of WIP is considered "healthy" in many factories. Cycle time is rarely tracked as a performance measure; it is even more unusual for a factory to make a conscientious effort to reduce it. Setup time may be targeted for improvement, but small-lot production is rarely pursued. Variability is often seen as an abstract concept when, in fact, it is responsible for most of our manufacturing inefficiencies.

Since the problems being solved by today's factory have not been recognized previously as being "problems," the solutions developed are often seen as unnecessary. If a factory transformation is going to be successful, the people in the factory need to understand what problems are being solved, why they are considered problems, and how they are being solved. Employees also need to know what roles they will play in the solution. They need to know that the principles are being applied and the practices are being implemented for good reasons.

Make the problem easy to see. Sell the problem. Clearly articulate *why* the transformation is necessary. This is a very effective way to motivate people to participate in the transition. It also instills a sense of urgency. Urgency should not be confused with anxiety. Urgency is critical for making the transition happen. Anxiety is not. We do not want to panic people into action; but without a sense of urgency, the transformation will not take place. Failure to establish this sense of urgency is the single largest mistake people make when trying to change an organization (Kotter 1996: 4).

When embarking on a factory transformation, the endings should be marked and even celebrated. An aerospace electronics manufacturer in southeastern Michigan recently had occasion to celebrate such an ending. The management wanted to reduce the customer lead time from order to delivery. Part of their solution involved establishing several single-piece flow work cells. (Recall from Chapter 5 that a single-piece flow work cell is the fastest way possible to process production material.) Once these work cells were operating, tens of thousands of pieces of WIP were no longer needed. Removal of the WIP was an *ending*, and it made many of the workers very uncomfortable! One night the manager (with the help of the authors) gift-wrapped one of the tower racks that had formerly held thousands of pieces of WIP worth many thousands of dollars. The "gift" was placed in the cafeteria for one week provoking dozens of guesses as to what it might be. Some people guessed a television set or a stereo system, and others thought it might be scrap! The management held an ice cream celebration to open the empty rack and commemorate *the ending of the need for the inventory.*

As supportive managers, our objective is to manage the factory transformation sensitively. It is important to understand who is losing something and, specifically, what they are losing. Regardless of whether the losses are real or perceived, objective or subjective, it is important to recognize that *people are experiencing an ending and that means giving up something.* The losses should be acknowledged openly and with genuine empathy.

It is very likely that workers and managers will exhibit signs of grieving and long for the good old days. For example, consider the transition required when an old pair of slippers is replaced with a new pair. Even if the old slippers were tattered and worn-out, putting on the new pair requires giving up something, even if it is just giving up that "old familiar feeling."

We should point out that it is not wise to criticize or denigrate the past in an attempt to get people to let go of it. People may have a fond nostalgia for the way things were, or they may have been a supporter or champion of the old way. Furthermore, the old way is probably part—possibly a large part—of their identity and may be a source of pride and feelings of accomplishment.

We have observed that some workers respond emotionally once they realize that the criteria on which their status, self-worth, and often-outstanding reputations are based are no longer valued. This can create anxiety and anger because the new measures and goals no longer recognize and reward the old criteria. We knew one woman who, for many years, was recognized by everyone in her factory as being the fastest assembler. She always had a large quantity of completed subassemblies at the end of the day, and these subassemblies would sit for days or weeks before someone needed them. When she was assigned to a work cell (table team), she had to coordinate her work with others who were slower, and she did not have her own stack of completed items at the end of the shift. The criteria on which her identity was based were no longer valued, and she had to make an "ending." It was more than she could take, so she took early retirement to the detriment of the company.

Belittling the past can cause deep resentment that is difficult to overcome later. It is better to present the new way as a development that builds on the solid foundation provided by the old way. The innovations and improvements in the future are made possible by the successes of the past. It is important to define and communicate clearly what has ended and what has not ended. The neutral zone is confusing enough without adding the confusion about what has actually ended.

## The Neutral Zone

The neutral zone is a time of great uncertainty and fear. It is like wading through a Florida swamp full of snakes, insects, alligators, and quicksand with no way to see very far ahead. By being aware of the hazards of the neutral zone, we can help the factory overcome them.

Bridges (1991) lists six dangers of the neutral zone.
1. Anxiety rises and motivation falls.
2. Absenteeism rises and productivity falls.
3. Old weaknesses and past resentments resurface.
4. Personnel are overloaded and turnover increases.
5. People become polarized.
6. The organization is vulnerable.

Now that the dangers are out in the open, it might seem like the neutral zone should be avoided at all costs! Perhaps it seems much safer to let the factory keep running the old way. However, like the "ending" described above, the neutral zone can be navigated safely by understanding the dangers and taking a few simple precautions.

The neutral zone is not simply a difficult phase that must be endured before the factory can operate successfully in its new configuration. The neutral zone is a beneficial period that provides the necessary time for reorientation and redefinition. People need time for the old methods and patterns of behavior to fade. They also need time for the new methods and patterns of behavior to take shape. As progress is made through the neutral zone, people begin to realize they wouldn't go back to the old way even if they could.

As managers, we must shield people from as much uncertainty as possible while in the neutral zone. We can do this by instituting temporary systems and policies, redefining reporting relationships, and setting measurable and achievable short-term goals. It is also important to give people the *opportunity to succeed*. If the goals are too ambitious, people will not achieve them and may begin to lose confidence. For example, the neutral zone is not the time to try to break productivity records. Also, people should be given the *tools to succeed*. It may be necessary to send workers and managers to special training. People may not understand fully what is going on around them, and they may feel forlorn. For this reason, it is important to strengthen interpersonal relationships and feelings of trust and loyalty while in the neutral zone. Top management should communicate regularly, reminding people why the change is necessary, where the company is going, how the company is doing. This information bears repeating over and over at every opportunity.

## The New Beginning

We mentioned earlier that transforming today's factory requires hard work. The new beginning is no exception. Often people are uneasy about the new way of doing things. The new beginning can revive some of the same anxieties that arose during the ending phase of the transition. After all, the new beginning is undeniable proof that the old way is finally gone. The new way may be seen as untried, causing more anxiety regarding whether the new way will actually work. People may be afraid that they will be punished if the new way does not produce the desired results. Furthermore, some people may thrive on the chaos, ambiguity, and lower expectations of the neutral zone. These people may demonstrate their objection to the new way by "goofing-off" and disrupting other people. Bridges points out that "for such people, the new beginning is an end to a pleasant holiday from accountability and pressure" (1991: 52).

The new beginning cannot be willed, forced, or announced into existence. Nelson observes that coercion is no longer an option; managers increasingly must serve as coaches to indirectly influence rather than demand desired behavior (1994). We must create an environment that encourages the new way to become established on its own. This is analogous to a seed being planted in the soil. No amount of wishing, planning, or announcing will cause the seed to sprout and develop into a healthy plant. However, by creating an environment that is conducive to plant growth, such as warm soil rich with nutrients and the proper amounts of moisture and sunshine, seeds will sprout readily on their own. Similarly, management should establish an environment that will encourage, support, and reinforce the new beginning.

There are four important ways management can nourish and sustain the new beginning (Bridges 1991: 52): explain the purpose, portray a vision, plan the transition, and involve the people.

### Explain the Purpose

The best way to encourage the new beginning is to explain the reason behind the transformation. If people understand the basic purpose of the transformation, they are much more willing to make it happen. If everyone understands the basic direction and goals of the company, there is a common basis for decision making. This leads to better decisions that are consistent with the overall business strategy. Furthermore, a stronger sense of cohesiveness is developed among the workers when they all understand the reason behind the transformation. The purpose should be communicated over and over again. It is essential to communicate the purpose clearly and often. It is often a good idea to use posters and banners so people are reminded constantly of the *purpose* of the transformation.

## Portray a Vision

Another excellent way to encourage the new beginning is to portray the "destination" in a way that engages the imagination. Like the vivid descriptions in an Edgar Allen Poe short story, we must depict the new beginning in such a way that people form a *mental image* of the transformed factory. Visual aids are useful for conveying an image of the new way. People want to understand the new spatial layout, how it will look, sound, feel, and even where they will be working. Something as simple as changing the softness of a table pad, moving a trash can, or providing workers with a new chair can cause tremendous distress if the change is not managed properly.

Sometimes it is useful for key managers and workers to tour a factory that has already made a similar transition to see how the outcome might look. Visiting such a facility is an opportunity for people to *see* the tools and techniques applied successfully as part of a coherent manufacturing system. This strengthens their resolve and helps the vision for their own factory seem within their grasp. By whatever means the image is conveyed, it is important that people begin to visualize it in their own imaginations.

Bridges (1991) offers two cautions about the visual image of the new situation. 1) People may be overwhelmed by the picture or unable to identify with it. If the vision does not seem real to people, or if they believe it is out of their reach, they will not be motivated to launch the new beginning. 2) People may believe that the picture itself has the power to make the transition happen. It is easy to dwell on the vision so much that it *seems* like it has been achieved even before the new beginning has been launched. If this happens, there will no longer be a perceived need to move ahead with the transformation, and the company will be stranded in the neutral zone.

The second caution is underscored by the 1998 *IndustryWeek* Census of Manufacturers (Jusko 1998c: 33-6). Corporate and plant executives were asked to identify manufacturing initiatives that they considered imperative if a company was to become "world class." The three initiatives identified are formal continuous-improvement programs, quality-management programs, and cycle-time reduction. Now let's look at three specific manufacturing strategies that support these initiatives: total quality management, lot-size reduction, and just-in-time

| Manufacturing Practice | Corporate Response | | Plant Response | |
|---|---|---|---|---|
| | Not Adopted | Widely Adopted | Not Adopted | Widely Adopted |
| Lot-size Reduction | 24 | 29 | 33 | 17 |
| JIT Production | 18 | 28 | 29 | 21 |
| Total Quality Management | 6 | 44 | 20 | 29 |

*Table 13.1* Adoption of Selected Manufacturing Strategies as Perceived by Executive Management

production. Table 13.1 shows the percentage of corporate and plant executives who believe the practice has either been widely adopted or not adopted at all in their companies.

It is interesting to note the disparity between the corporate and plant responses. For example, consider total quality management (TQM): 44% of the corporate executives surveyed believe that their companies have widely adopted TQM. In contrast, only 29% of the plant executives feel they have widely adopted TQM. 29% of corporate managers believe their companies have widely adopted lot-size reduction, while only 17% of plant managers agree. The contrast is even more significant when we consider the "not adopted" responses. 18% of corporate executives feel their companies have not adopted JIT production, while 29% of plant executives admit they have not adopted JIT. Only 6% of corporate executives feel they have not adopted TQM, but 20% of plant executives say they have not adopted TQM.

So, the board members have one perception of reality, and the factory managers have quite another. One explanation for the disparity is the degree to which corporate executives are insulated from the shop floor. Thinking about making a change or even deciding to make a change is not the same as making a change. Similarly, developing a vision and making that vision a reality are also very different. Once the vision for the future has been established and communicated company-wide, it is easy to forget that time and effort are still required in order to actually transform the factory. Manufacturing managers are usually much closer to the actual plant operations and may be less susceptible to this common delusion.

### Plan the Transition

The launch of the new beginning should be supported by a step-by-step transformation plan. One obvious reason for a transformation plan is project management benefits that come from planning any activity. *The transformation plan also serves an important psychological role.* The existence of the plan sends a message loud and clear: somebody is looking after us, taking our needs seriously, and watching out so we don't get lost along the way. A transformation plan is much more than a future state diagram or a list of tasks for getting the factory from here to there. It should address how and when people will receive information, training, and support, and how people can provide their own input to the transformation process. A successful transformation should also identify new values, behaviors, and attitudes that are appropriate and how they will be instilled and encouraged. A transformation plan should focus on the transition process, not the outcome.

### Involve the People

People need to know where they are going. They need to understand their future role. For human beings, uncertainty can be exhilarating, but in matters of livelihood, it can also be upsetting and even frightening. Most people want to know what they will be doing, how they will interact with others, and how their interpersonal relationships will change after the factory has been transformed. This alleviates fear and promotes trust, understanding, and commitment to the transformation.

People also need to participate in the transformation process itself. There are two important reasons for this. 1) Involving everyone ensures that everyone understands the "big picture" goals and the reasons for the transformation. This is important for aligning behavior with the corporate goals and establishing a shared value system. 2) Shop floor workers represent a wealth of practical knowledge and information. Their participation promotes ownership in the process and commitment to the outcome.

## Resisting Change

Change is almost always resisted to some degree. By understanding why change is resisted, we can mitigate the resistance and make the change process more enjoyable (and less painful). Below are ten common reasons change is resisted.

1. Loss of control: Doing *to* us rather than *with* us, not involving people affected in the planning process.
2. Uncertainty: Management not portraying confidence in or commitment to the new way.
3. Lack of information: Secrets or surprises.
4. Too different: Unfamiliarity.
5. Loss of "face": Embarrassment over having supported the old way.
6. Insufficient knowledge: Concerns about future competence.
7. Ripple effects: Disruption and fear.
8. Personal loss: Position, job, power, perks.

9. More work: Longer hours, harder work.
10. Past resentment: Bitterness over old injustices.

## *Rules of Transition Management*

There are four rules of transition management that are especially appropriate to factory transitions: *be consistent, ensure quick success, symbolize the new identity,* and *celebrate the success* (Bridges 1991: 61-63).

### Be Consistent

This is often more difficult than it seems. Specifically, we must guard against three types of inconsistency:

* inconsistency of messages (sending mixed messages)
* inconsistency of actions (saying one thing and doing another)
* inconsistency of rewards (demanding one thing and rewarding another).

In our experience, the third type is by far the most common form of inconsistency in a factory transformation. For example, after the transition many of the old performance measures are no longer appropriate. Continuing to reward people based on the old measures encourages them to disregard the new way and cling to the old. *Old measures may not reflect the progress made toward new goals.*

### Ensure Quick Success

A quick success bolsters confidence in two ways. First, people are reassured; the confusion and low productivity often associated with the neutral zone can cause people to lose confidence in themselves. They may begin to doubt their own abilities. A quick success can restore morale and nurture the new beginning. Second, people gain confidence in the new way.

Just as it might take several months for a seed to germinate, grow, bloom, and eventually bear fruit, it could take several months for the new beginning to truly become established and begin providing the promised benefits. A quick success gives people a chance to see for themselves how well the new way works. The importance of short-term success is emphasized by Kotter: "Without short-term wins, too many employees give up or actively join the resistance" (1996: 11).

### Symbolize the New Identity

This is both a rule and a caution. During a transformation, people attach meaning to almost *everything* that happens. The actions of managers and change agents are scrutinized closely by people who are looking for hidden symbolism. Anything we say or do can be interpreted symbolically, so we must establish the right symbols. Usually we think of symbols as objects rather than actions. An excellent way to establish symbols is to use icons (objects, posters, photographs, floor layouts, charts showing progress toward new goals, and banners) to remind people where the factory is going and how it is going to get there. One office furniture factory in southern Michigan has a full-size team kayak hanging from the ceiling in the cafeteria to symbolize the transition to team-based manufacturing.

### Celebrate the Success

Just as the ending should be marked and even celebrated, the new beginning should be marked and celebrated as well. It is not entirely clear when and how often the celebrations should take place. After all, the speed with which people accept the new beginning varies greatly from one individual to the next. Furthermore, a full transformation that permeates the corporate culture can take from three to ten years! Our suggestion is to celebrate early and often rather than waiting several months or years for the transformation to be well under way. Management can use celebrations to publicize the early successes, reinforce desired behavior, and remind people what is valued in the new scheme of things. The celebrations themselves can even become a symbol of the new beginning. The celebrations can be relatively simple, such as the managers passing out ice cream bars at break time, or providing deli sandwiches for members of new work cells to celebrate their first week.

Early celebrations reinforce the new beginning. On the other hand, a "victory celebration" should not be held too soon. People will know it is too soon, and managers will lose credibility. Furthermore, people may misinterpret the "victory celebration" as a lowering of expectations. Kotter (1996: 13) warns, "The premature victory celebration stops all momentum."

Some researchers warn that it is dangerous to look back and say, "See how far we have come." We disagree. Tracking and publicizing improvements throughout the transformation process reminds *everyone* that progress is being made. This is motivating and encouraging. It is amazing how fast people forget how the factory really was before the transformation. After the changes have started, it is difficult to think of the factory as having been any other way. People have selective memories, and the "old days" miraculously become the "good ol' days."

# Management Commitment

Transforming today's factory requires the involvement of everyone in the organization, as well as the total commitment and dedication of top management. It requires taking a good, hard look at the factory to see how it really runs (assessing objectively the factory operations). It may require doing some things differently (realigning the management strategy to be compatible with corporate goals) (Tucker and Davis 1993: 60).

In today's factory, top managers, by their policies, decisions, and allocation of resources, should see themselves as supporting the production workers. In a sense, the managers, executives, and office workers become the support staff of the production workers. The organizational structure is turned *upside-down*. The workers, in turn, have much greater authority and responsibility for production matters (Buzacott 1995: 122). This requires top managers to relegate much of the decision making and problem solving to lower level managers and production workers. Workers at all levels need to be trained to handle problems immediately. In today's factory, problems should be handled by the person (with the capability) who is closest to the problem.

Successful decentralization requires a company-wide understanding of the business goals. This ensures that the solutions implemented and the decisions made are in alignment with the overall corporate strategy. Schonberger states, "World-class manufacturing requires that *everyone* help manage the enterprise, that all employees be involved up to their ears in the pursuit of continual and rapid improvement" (1986: 217). Employees at all levels should be taught the principles and practices for running today's factory.

Top management commitment will be tested almost as soon as the factory transformation begins. For example, manufacturing managers tend to read and rely on the same performance reports week after week. These reports usually reflect short-term financial measures such as direct-labor productivity or standard labor variance. Successful managers understand the reward or penalty system associated with these performance measures and the importance of receiving favorable reports. Unfortunately, most constructive change causes the short-term measures to decline. Top management should be willing to 1) accept some temporary performance degradation as measured by previous reports, and 2) support new reports that reflect new performance measures and new goals. Otherwise, it may be in a manager's *personal* best interest to maintain the *status quo* (Ashton, Fagan, and Cook 1990: 86).

How important is top management commitment? Top management commitment is absolutely essential, for without it, the transformation process will surely fail. "Ultimately, lean systems can provide American industry the ability to thrive in the increasingly competitive global environment ahead. That's the prize: the future. But it all starts—or ends—with the commitment of the person at the top" (Day 1994: 70).

How important is middle management commitment? Middle management commitment is also essential. Ideally, top management should expect and demand that every member of the management team understands the principles for running today's factory and is in full support of the transformation (Ashton, Fagan, and Cook 1990: 87).

Regardless of whether the company is a small family-owned business or a Fortune 500 corporation, the middle managers are almost always the most resistant to change. "Those managers who simply will not accept the new ideas will have to be removed. This may sound harsh, but managers who will not cooperate cannot be allowed to compromise the conversion process" (Bergstrom 1995: 33).

Running today's factory requires a unified effort with every member of the company working toward a common corporate goal. Unfortunately, a small percentage of managers will never accept the new order (Womack and Jones 1996: 259). If left in place, even a small group of managers can possess enough inertia to slow, stagnate, or even stop a lean transformation. We have found that through education, coaching, and an appropriate performance appraisal system, discharging managers is usually unwarranted. Nevertheless, top management must be willing to take whatever steps are necessary to ensure the success of the factory.

Here are some suggestions for top management.

- Be the visible champion of the transformation process!
- Communicate the corporate goals to all employees.
- Develop a united management team to lead the transformation.
- Teach all employees the principles for running today's factory.
- Train all employees in problem solving.
- Decentralize decision making and problem solving.
- Accept some short-term degradation of familiar performance measures.
- Adopt a supportive attitude.

Transforming today's factory may be necessary, but it can also be uncomfortable and even frightening for almost everyone. Recalling the upside-down organization described earlier, management must take an almost *ministerial* role and ease the concerns and fears. That is why management must prepare everyone mentally and culturally for the transformation (Inman and Brandon 1992: 58).

## Company Culture

Many factory managers lament their existing company culture and express a need for an urgent culture change. Dirk Jager, the new CEO of Procter & Gamble, promises to "sweep away the *cultural cobwebs*" at the consumer products giant. Jacques Nasser, the new CEO of Ford, promises a "*major cultural overhaul.*" Furthermore, Nasser actually tells managers who are resisting change that "if they can't get with the program, they don't deserve a job" (Lancaster 1999). What exactly do these leaders of industry mean by the term *culture*? What is a *cultural cobweb*? Can a culture be *overhauled*? The word *culture* is a loosely used term in the American manufacturing community, and the specific meaning isn't always clear. Here we explain the relevance of company culture to running and transforming today's factory. We define clearly what is meant by *corporate culture* from an anthropological point of view. We also provide insight for modifying and aligning the existing culture to complement new corporate strategies.

There is a growing body of research indicating that 1) corporate culture exists, and 2) understanding and shaping the culture to fit the company goals can be a significant competitive advantage (Frost 1994). Corporate culture can be modified to improve quality, productivity, profitability, and the work environment in general (Schein 1986; Amormino 1992). Furthermore, Barney (1986) even suggests that superior corporate performance is a reflection of superior organizational culture.

Culture is defined as a shared pattern of beliefs and behaviors common to a group of people. The group can be any size from an extended family to a nation. The group identity does not necessarily have to be geographic or national; it can be religious, social, political, recreational, or occupational. In fact, occupational identification in some fields may be stronger than many other forms of identification. For example, teachers, engineers, or computer programmers tend to share many beliefs and behaviors with others in their occupational group regardless of their regional or national affiliations. This group identification is obvious to attendees at a political, religious, computer peripheral, engineering, or anthropology convention.

The important point is that culture is shared and learned, and it is not something that can be superficially imposed. For many years, engineers were stereotyped as wearing horn-rimmed glasses and pocket protectors. In fact, many still do! Nevertheless, the "engineering culture" goes much deeper than narrow ties, white shirts, and white socks.

Shared beliefs and behaviors are based on underlying assumptions about the physical and supernatural worlds and the nature of human relationships. The underlying assumptions reflect the human attempt to understand existence and to explain the observable and spiritual universe. The underlying assumptions and resulting values are much more difficult to discern, but they can be deduced by observing people's behavior. They help us answer questions such as why is the world the way it is? Why are we here? How large is the universe? Where did we come from? It is difficult to bring these underlying assumptions into focus. They are rarely articulated or even acknowledged; yet, they are largely responsible for the way we perceive, think, and feel in response to circumstances we encounter in our daily lives. It is the underlying assumptions that the anthropologist seeks in order to understand, define, and explain a culture.

These underlying assumptions can be quite different from culture to culture. Take a moment and consider your own assumptions about the nature of the universe. Now, consider the view held by the Lacandon Maya, a small group of Maya living in *La Selva Lacandona* in Chiapas, Mexico.

For the Lacandons, levels of existence are stacked like dinner plates (or tortillas), and human beings are able to travel from one level to the next if they are especially devout (or dead). This world, the realm of humans, is the second level in the Lacandon universe. The sky is the underside of the third level, which is occupied by a pantheon of gods. The gods are greedy and bloodthirsty with a taste for human flesh, babies being especially delicious! They constantly demand offerings of food, incense, cigars, and other seasonal items. If the gods are not satisfied, they retaliate by making people sick. The lowest level, the underworld, is ruled by a malevolent supernatural figure, *Kisin*, who punishes people after they die for offenses that include theft, murder, incest, and lying. When the punishment is finished, the person ascends to the third level to begin performing everyday tasks for the gods. Interestingly, the Mayan term for "the world" is *toworoh kash*, which means literally the entire forest! Until the past decade, the Lacandons were isolated in the vast rain forest of northern Chiapas, and the forest truly was their whole world.

Values and expressed behavior are a natural outgrowth of the underlying assumptions. The premises on which the Lacandon Maya base their lives might seem unusual. Yet, given the underlying assumptions about their universe, their values and behaviors are appropriate and understandable. Observable behaviors that are acceptable include truthfulness, respect for others, being a successful farmer (so the individual will have offerings to give to the gods), and diligence in ceremonial observation. Behaviors that are not observed or rewarded in the Lacandon culture are generosity, charity, affection, and taking care of one's parents. These underlying assumptions are internalized and corresponding values are established in childhood.

There are fewer Lacandons (approximately 250) living in Chiapas today than there are employees in many factories. Yet, the Lacandon culture is one of the most distinct in all of Middle America. So, it is the strength of identity and the sharing of values and beliefs that defines a specific culture and gives it character, not the size of the group.

Just as a culture is characterized by the pattern of professed beliefs and observed behavior and interactions, a company culture is characterized by the pattern of professed beliefs and observed behavior shared by employees of the company. Job descriptions, policies, expectations, actions that are acceptable, actions that are prohibited, how people perceive problems, how they find solutions, who makes decisions, how people interact with co-workers, supervisors, customers, and suppliers—all are visible indications of the company culture.

These are the observable behaviors that are, in turn, engendered by the underlying assumptions and values of the company. The rules and observable behaviors are indications that there is a deeper set of assumptions about the nature of the "work" in that particular company. Each company has a unique culture, and the successful employee can discern and follow the rules.

Anyone who has worked for more than one company knows from experience that different companies have different work ethics, social relationships, and acceptable or unacceptable conduct. There may be formal communication, such as an employee handbook, that instructs employees about some aspects of the company culture, but the communication is more often implicit and subtle. For example, there may be a formal dress code, but there are also *unspoken rules*, such as "no turtlenecks" or "no tennis shoes." In some companies, there is an *implied* arrival time of 7:00 A.M. even though the official start time is 8:00 A.M. Turning off the restroom light can be "expected" in some factories. Sometimes status is indicated by the type of office chair (arms or no arms, low back or high back, leather or cloth) or whether or not the office has a door. In some companies, the size of the office or cubicle (in square feet) directly reflects the status of the occupant (a different size workspace for secretaries, engineers, supervisors, managers, vice presidents, and so forth).

## Forms of Company Culture

There are two classifications of company culture that are relevant to our discussion about today's factory: hierarchies and clans. We prefer these terms because they are consistent with the scientific literature on this subject. Hierarchical and clan cultures represent contrasting degrees of organizational complexity; they are not *discrete* types but are the extremes of a continuum. Every company culture exhibits characteristics of both and can be placed somewhere along the continuum. A company does not have a homogeneous clan-based culture or a purely hierarchical culture. Nevertheless, using this mental construct can be useful in understanding the implications of company culture for running today's factory.

*A hierarchical culture is characterized by a formal structure with individual roles, responsibilities, and limits of authority.* Job descriptions are narrowly defined, and people are preoccupied with their own duties. In a hierarchical culture, control is imposed through the formal chain of command. Orders come from the centralized authority figure, and compliance is ensured by direct supervision. Just as a hierarchical organizational structure is more appropriate for companies in a stable environment, a hierarchical culture is appropriate in factories where the manufacturing systems are relatively stable over time. Why? Companies with stable manufacturing systems have less need to adapt, innovate, change, and restructure their production units.

In contrast to hierarchical culture, *a clan-based culture is characterized by ever-widening circles of loyalty, cooperation, and mutual support.* In a clan-based culture, control is maintained through guidance and support. The clan-oriented culture is conducive to change and innovation. It is more responsive to outside influences and fosters interaction among other departments, suppliers, and customers. For these reasons, a clan-type culture is ideal for a factory that exists in an uncertain or rapidly changing environment.

Company culture is not homogeneous, and, in fact, some departments in a company may exhibit one culture, while others exhibit a different culture. In such a case, each department would be a subculture of the overall company culture. The heterogeneity of company culture can be observed in companies where functional departments are spatially separated.

To see this for yourself, adopt the role of an anthropologist and make some objective observations about your factory. Stroll briefly through several departments such as accounting, information technology, engineering, and production. During your stroll notice how the space is arranged. How are the walls and dividers placed? Are some areas carpeted? Are some areas air-conditioned? Are people grouped together? How close together are they? Are common areas available where impromptu meetings can be held? Is a formal conference area nearby? Are departmental goals displayed? Is "the boss" or manager singled out? How are work surfaces organized? Which personal items, if any, do people display at their work areas? What are people wearing: casual or dress, company uniform? Is the dress different between labor and management? What are people doing: working alone, working in teams, talking, joking, meeting, wandering, daydreaming, or surfing the web? What do you hear: music, talking, laughing, telephones ringing, machinery running, or the intercom?

What do the answers to these questions tell you about the values and underlying assumptions on which your factory's culture is based? Do different departments exhibit different attitudes and behaviors? The chances are that the behaviors and values exhibited by the different departments or areas visited are distinctive. As we indicated above, occupational culture may be more easily recognizable than national culture. Nevertheless, at some level, there is a shared value system associated with the various subcultures in an organization. If people cannot adapt and become acculturated into the company, they are usually very uncomfortable. They often leave to find a company that is more compatible with their personal beliefs.

# Corporate Culture and Manufacturing Strategy

What type of company culture is most appropriate for today's factory? What values and beliefs promote a lean operation? What kinds of behavior support the corporate goals? We have hinted strongly that a clan-based culture is preferred because it adapts more readily to the changing conditions that today's factory faces. However, we need a more definitive answer. Beyond being adaptive the company culture must support a sound manufacturing strategy that allows the company to reach its goals. This issue was recently addressed in a scientific study of 138 manufacturing companies. The sample included traditional American-owned companies, progressive American-owned companies, and Japanese-owned companies located in the United States. The following statement summarizes the results.

> A manufacturing strategy which is formalized, communicated, long-range oriented, linked to business strategy, and intended to create competitive manufacturing capability is likely to reside in an organizational environment characterized by coordinated decision making, use of small groups and teams, decentralized authority, high employee loyalty, and a shared plant-wide philosophy. Without speculating about causality, the data indicate that a well aligned and implemented strategy exists in plants where common goals rather than hierarchy is the dominant control mechanism (Bates, Amundson, Schroeder, and Morris 1995: 1574).

This particular study investigated correlation but not causality. The data establish clearly that a strategy that is well aligned with company goals is more likely to be found in a company with a clan-based culture. However, the researchers did not investigate whether the clan-based culture is actually the cause of the manufacturing strategy being so well aligned with the company goals. Regarding the correlation there are three possible explanations.

1. A clear corporate purpose engenders loyalty and strengthens shared values.
2. A clan-based culture makes it easier to establish an overall manufacturing strategy.
3. It is mere coincidence that clan-based culture is present in companies with well-aligned strategies and goals.

Our experience suggests that clan-based culture and manufacturing strategy reinforce one another (explanations 1 and 2). Furthermore, we believe a clan-based culture actually enables the successful implementation of a well-aligned manufacturing strategy.

At the beginning of a factory transformation, managers need to evaluate the existing company culture. Understanding what behaviors and attitudes the current culture encourages gives us clues about the values and underlying assumptions. This is one way managers can identify positive values that may be missing at present, and identify negative values that need to be purged.

We are familiar with many manufacturing companies that operate on the underlying assumption that "more" and "faster" are better on the assembly line. For one company, in particular, it is interesting that this underlying assumption has endured, since it has been many years since anyone was evaluated, rewarded, promoted, given a raise, or fired based on daily production rates. Nevertheless, some of the long-term employees still assume that they should continue to build product regardless of whether the next process is ready for the material. The unspoken "value" reflects an admirable work ethic of responsibility and diligence but, at the same time, it is inconsistent with the lean production flow that is in place today. The old value leads to excess and potentially defective inventory accumulating between processes. This buildup causes cycle time to increase and exposes the factory to financial risk.

## Culture Change

How can a company change its values and underlying assumptions? One thing is certain: underlying assumptions cannot be modified by direct attack. Underlying assumptions have been internalized, and the resulting values are largely subconscious. Most people don't even realize they have them. *Telling* people that their innermost convictions are no longer valid or valued will not persuade them to change their beliefs. It *can* make them resistant, defensive, and angry. The previous culture can and should be respected even though it is no longer encouraged. The underlying assumptions of a company culture are developed over time by working with other people and observing interaction in the company and by participating in its system of reward and punishment. Mere words cannot change values and underlying assumptions that have been reinforced through years of observation, interpretation, and internalization.

Instead, management can determine what types of behaviors are required in the new order and then promote those behaviors. Actions can be modified relatively easily through communication, education, support, rewards, and reprimands. *When people begin to perceive the new behaviors as normal rather than unusual, the company has made its first step toward a cultural revolution.*

As benefits to the factory and to the individual become apparent, residual skepticism and resistance begin to fade. These benefits can include: a steady work flow; the opportunity to learn new skills; a safer, cleaner, and more enjoyable work environment; increased respect for others and oneself; and a higher salary due to the increased profitability of the factory. Gradually people become aware of the cause-and-effect relationship between the new behaviors and the benefits. Most people's values undergo a transition, and in time the underlying assumptions may even begin to change. This, however, occurs slowly and even subconsciously as a result of steady exposure to the new way of doing things in today's factory.

## Organizational Structure to Support a Transformation

Just as there are no generic solutions for running today's factory, there are also no generic structures for organizing today's factory. There are, however, guidelines for choosing an appropriate organizational structure, corporate culture, and management style. The key to organizational design is to acknowledge that the factory does not exist in isolation. A factory undergoing transformation exists in a complex and unstable environment, and the organizational structure should be designed accordingly.

As we have seen throughout this book, many of the principles and practices for running today's factory differ rather significantly from traditional manufacturing wisdom. Some, such as *frequent changeover* and *small-lot production*, contradict mass production tenets. Even though lean techniques are becoming more familiar in American manufacturing, *transforming the factory still requires radical changes in the way people think about manufacturing.*

Organizational structure can enable or inhibit a company's ability to change. Since transformation is a radical change, the organizational structure of the factory may need to be altered to support a successful transformation. Such a change to the organizational structure can be considered a step toward factory-wide continual improvement. Over time, the organizational structure may need to evolve as the environment changes. *Because a factory transformation is a radical change, it should be no surprise that the structure that is in place at the initiation of a transformation is not necessarily the same structure that can sustain the transformation after it is underway.*

### Initiating the Transformation

Strong focus, direction, and visible support from *a powerful person or guiding coalition may be the only way to initiate successfully the transformation process.* This centralization of authority and leadership may appear to contradict our previous discussions about giving authority and skills to workers to make decisions and solve problems. The reason for this centralization, however, is quite clear. Initiating

263

such radically different manufacturing strategies requires a powerful, centralized coalition of authority to allocate resources, express support, suppress opposition, assure workers that the new way is permissible, and even reassign those who cannot accept the new order and would hinder the transformation. The most important reason for having centralized authority is that the message received by all employees is clear and consistent and the goals are coherent.

Another structural element that influences a company's ability to initiate successfully a period of innovation is formalization. Formal organizational structure discourages deviation from existing procedures, whereas *informal organizational structure promotes innovation and the creation of new procedures.* An informal structure essentially allows people to learn. It provides people with the freedom to experiment, take risks, and initiate the new way of doing things without fear of reprisal.

When initiating a factory transformation, it is important to maximize the flexibility of the workers both in their job skills and in their willingness to perform new tasks. When jobs are highly specialized, people tend to pay excessive attention to their individual tasks. This does not promote system-wide optimization. When workers are cross-trained, they are exposed to a wide spectrum of processes and activities. They can gain an understanding of how a change in one part of the system affects the entire factory.

Job rotation is an excellent way to maximize job flexibility (or minimize job specialization). We are currently working with a manufacturer who has successfully initiated six manufacturing cells in one product family. One cell was particularly diligent about rotating the workers from one job to the next throughout the day. Soon, every member of the team could perform all the various jobs in the cell. Later, one of the team members was promoted to department supervisor. She initiated the same type of job rotation for every job in the department. This not only established a sense of teamwork but also an understanding of the problems each job brings. After this department had rotated jobs for several weeks, one woman told us she would never complain again because she learned that even if a job looks easy it might not be.

Job rotation helps each worker understand the impact his or her work has on other workers at downstream processing stations. It also gives the department flexibility to cope with uncertainty. Of course, not everyone will become an expert at every job on the rotation schedule, but everyone will be able to do many jobs if necessary. We have heard marketing people lament that their urgent "rush orders" could not be produced because a particular operator was absent. Cross-training workers allows the department to produce regardless of absenteeism because each worker can step in and perform other jobs. We should point out that such rotation could cause productivity to drop slightly as each worker learns new skills. However, the long-term strength and flexibility of the department is greatly enhanced.

The transformation process is much easier to initiate in companies with fewer organizational layers. This is primarily because information can be communicated more freely. Clear, timely, and uncontaminated information becomes increasingly important when a transformation is initiated; a lot is changing, and there is important information about jobs, schedules, relationships, and policies that needs to be communicated to the right people.

During the transformation, some people will not fully support the transformation effort because they feel their status or position or other personal interest is at stake. These employees may block or distort information; this is more likely if there are more levels in the organizational hierarchy. Companies with fewer organizational layers experience less contamination and smoother flow of information throughout the factory (Koufteros and Vonderembse 1998: 2871).

## Sustaining the Transformation

Just as centralized authority is necessary when initiating a factory's transformation, *a powerful, centralized coalition is also required to sustain the transformation.* Again, this does not contradict the idea of decentralized problem solving and decision making. In fact, the process of relegating authority to lower levels of management and to shop floor workers requires the full support and commitment of top-level management. For example, centralized strategic planning produces unified corporate goals. When these goals are communicated throughout the company, it is possible for every worker in the company to understand the goals, make the right decisions, and select appropriate courses of action. In this way everyone participates in the new order.

The transformation will undoubtedly require the reallocation of human and technical resources. For example, equipment that has been in a functional department for many years may be relocated in a multifunctional work cell that supports a specific product family. Job assignments and reporting relationships will change, and entire departments may be dismantled. A powerful guiding coalition keeps people focused on the overall goals and provides a way for people to resolve any conflicts. Furthermore, a centralized authority can assemble and allocate sufficient resources wherever they are needed to ensure success.

*Sustaining the transformation requires a formalized organizational structure.* Loose informality during the *initiation stage* promotes necessary experimentation and risk-taking; however, an informal structure during the *sustaining phase* leads to confusion, conflict, and ambiguity. If the organizational structure is formalized and guidelines are established, people will understand clearly how the transformation should proceed and how their job activities have changed. Without these structural guidelines, people may tinker and experiment with their new job assignments. While continual improvement is an objective, it

265

should be balanced by standardization (Chapter 7). Tampering too soon with the new methods may just be an excuse to revert to the old way of doing things.

*Successfully sustaining the transformation may require a higher degree of job specialization.* This is especially true of middle management roles. Consider a factory that is abandoning its familiar MRP scheduling system in favor of a pull system (Chapter 9). If the individual formerly responsible for the MRP system (possibly the materials manager) is also going to be responsible for the pull system, he or she will need to acquire a considerable amount of specialized knowledge. (While there is nothing difficult about operating a pull system, there are some intricacies involved in establishing one. Pull systems require very thoughtful planning and understanding of the factory, the value stream, and the principles of pull.) It may not be feasible to rotate people through this kind of job until the transformation has been underway for quite some time.

During a factory transformation, people's jobs and responsibilities change. If new job responsibilities are added too early, or if job rotation is introduced too quickly, people may feel overwhelmed. In general, by focusing on specific tasks, new jobs can be learned faster and performed with more competence and confidence. After the transformation is well underway, experimentation, cross-training, and job rotation can begin. Eventually job flexibility will become the norm.

*An organizational structure with few levels helps sustain the transformation.* Again, this is primarily because a successful transformation requires the smooth, efficient, and undistorted flow of information. Information can be communicated most efficiently in an organization with few levels in the hierarchy. Organizations undergoing a transformation should promote integration among functional groups. This includes coordinator roles and extensive use of cross-functional teams (Daft 1992: 178-88). Table 13.2 summarizes structural elements that are particularly relevant during the transformation of today's factory.

| Structural Element | Initiating a Transformation | Sustaining a Transformation |
|---|---|---|
| Centralization | Centralized authority | Centralized authority |
| Formalization | Informal | Formal |
| Specialization | Non-specialized | Specialized |
| Communication | Top-down | Cross-functional |
| Organization | Few hierarchical levels | Few hierarchical levels |

*Table 13.2 Summary of Structural Elements to Support a Transformation*

# Summary

Throughout this book we have emphasized that there are no generic solutions for today's factory. Managing the transformation process is no different. In every factory, the people, processes, culture, and organizational structure are different. Nevertheless, we have developed a framework that can be useful for planning and organizing the transformation, and it begins with understanding the business goals. We feel strongly that running today's factory successfully requires first understanding the overall business strategy.

Top management commitment is the single most important element of a factory transformation. Success depends on having a sufficiently powerful guiding coalition to lead the transformation. In this chapter we present several suggestions for top management that can lead to a unified effort with every member of the company working toward the common corporate goal. Transforming today's factory is not easy, but by managing the transition carefully, it can be a time of revitalization and innovation.

Any successful transition begins with an ending, that is, an ending to the old way of doing things. People may experience a sense of loss, confusion, and anger when they have to relinquish their established roles. Fortunately, there are clear guidelines for managing transition effectively through the neutral zone and into the celebrated beginning.

Each factory has its own unique culture, and its people share certain fundamental assumptions and deeply ingrained values that influence how they go about their work. When a factory undertakes a transformation, the people who work there must change also. It is easier and faster to change people's behavior than to change their values and beliefs; these more profound changes can take place after the new way becomes the norm and the benefits of the transformation become evident.

Organizational structure is an important concern during the transformation of a factory; it can enable or inhibit a company's ability to change. The organizational structure of the company may need to change in order to support the transformation process. The transformation process is easier in companies with few organizational levels. These companies have a clan-type culture characterized by common goals, cooperation, and mutual support. Communication is especially important during the transformation, and it must be timely and truthful. The informal, interactive environment of the clan-type company is conducive to change and innovation; it is also indicative of an organization having well-aligned manufacturing and business strategies.

"There are no experts, just people with more experience. The longer we wait, the more experience our competitors will have when we start" (Shook, *in* Liker, 1997: 69).

The principles presented here are intended to help *keep the factory running on a steady course today and tomorrow.*

# Bibliography

**Adam, E. E.**
1994 Alternative quality improvement practices and organization performance. *Journal of Operations Management* 12(1):27.

**Adams, William G.**
1999 Interview by authors. Haines City, FL, 6 January.

**Aldred, K.**
1998a Survey reveals confidence in flow manufacturing techniques. *IIE Solutions* 30(4):8-10.
1998b Experts warn manufacturers not to chase low wages across the globe. *IIE Solutions* 30(8):6.

**Alwan, L. C.**
1991 Autocorrelation: Fixed versus variable control limits. *Quality Engineering* 4(2):167-88.

**Amormino, M.**
1992 Hidden talents: How to take the lead with employee involvement teams (Part 3). *Plant Engineering & Maintenance* 15(1):27-33.

**Anderson, J., R. Schroeder, S. Tupy and E. White**
1982 Material requirements planning systems: The state-of-the-art. *Production and Inventory Management* 23(4):51-67.

**Arkansas Economic Development Commission**
1998 Arkansas Fact Finder.

**Ashton, J. E., R. L. Fagan and F. X. Cook**
1990 From status quo to continuous improvement: The management process. *Manufacturing Review* 3(2):85-90.

**Askin, R. G. and C. R. Standridge**
1993 *Modeling and Analysis*. New York: Wiley.

**Barney, J. B.**
1986 Organizational culture: Can it be a source of sustained competitive advantage? *The Academy of Management Review* 11(3):656-65.

**Bates, K. A., S. D. Amundson, R. G. Schroeder and W. T. Morris**
1995 The crucial interrelationship between manufacturing strategy and organizational culture. *Management Science* 4(10):1565-80.

**Bergstrom, R. Y.**
1995 Lean principles & practices. *Production* 107(8):32-33.

**Blood, B. E.**
1994 Read my lips—no more late deliveries. *Hospital Materiel Management Quarterly* 15(4):53-55.

**Bornholdt, O. C.**
1913 Continuous manufacturing by placing machines in accordance with sequence of operations. *Journal of the American Society of Mechanical Engineers* 35:1671-78.

**Bowman, R. A.**
1994 Inventory: The opportunity cost of quality. *IIE Transactions* 26(3):40-47.

**Bridges, W.**
1991 *Managing Transitions: Making the Most of Change.* Reading, MA: Addison-Wesley Publishing Company.

**Buchanan, J. and J. Scott**
1992 Vehicle utilization at Bay of Plenty Electricity. *Interfaces* 22(2):28-35.

**Buzacott, J. A.**
1995 A perspective on new paradigms in manufacturing. *Journal of Manufacturing Systems* 14(2):118-25.

**Buzacott, J. A. and J. G. Shanthikumar**
1993 *Stochastic Models of Manufacturing Systems.* Englewood Cliffs, NJ: Prentice-Hall.

**Carlier, J. and E. Pinson**
1988 An algorithm for solving the job-shop problem. *Management Science* 35(2): 164-76.

**Carlton, J.**
1997 Packard Bell plans direct sales of its PCs to business customers. *The Wall Street Journal,* 18 June.

**Chase, R. B. and D. M. Stewart**
1994 Make your service fail-safe. *Sloan Management Review* 35(3):35-44.

**Choi, T. Y. and K. Eboch**
1998 The TQM paradox: Relations among TQM practices, plant performance, and customer satisfaction. *Journal of Operations Management* 17(1):59-75.

**Chowdhury, S.**
1998 One on one with Crosby. *Automotive Excellence* 1(Winter):12-16.

**Clark, L.**
1953 *The Rivers Ran East.* New York: Funk & Wagnalls.

**Clinton, B. D. and K. Hsu**
1997 JIT and the balanced scorecard: Linking manufacturing control to management control. *Management Accounting* 79(3):18-24.

**Cole, R. E.**
1992 The quality revolution. *Production and Operations Management* 1(1):118-20.

**Coleman, C. Y.**
1998 Pushing paper in a plastic world. *The New York Times,* 23 February.

**Cook, D. P.**
1994 A simulation comparison of traditional, JIT, and TOC manufacturing systems in a flow shop with bottlenecks. *Production and Inventory Management Journal* 35(1):73-78.

**Cooney, J.**
1995 Review of Fad surfing in the boardroom, by E. C. Shapiro. *Business Quarterly* 60(Winter):95-96.

**Crawford, K. M. and J. F. Cox**
1991 Addressing manufacturing problems through the implementation of just-in-time. *Production and Inventory Management Journal* 32(1):33-36.

**Daft, R. L.**
1992 *Organizational Theory and Design.* 4th ed. Saint Paul: West Publishing Company.

**Daniel, S. J. and W. D. Reitsperger**
1996 Linking JIT strategies and control systems: A comparison of the United States and Japan. *The International Executive* 38(1):95-121.

**Davis, D.**
1978 Ritual of the Northern Lacandon Maya. Ph.D. diss., Tulane University.

**Davis, D. and C. T. Standard**
1997 Celestial phenomena in Lacandon Maya song and lore. In M. Preuss, ed., *LAILA: Beyond Indigenous Voices,* pp. 47-51. Culver City, CA: Labyrinthos.

**Day, J. C.1994**
The lean-production imperative. *IndustryWeek* 243(15):70.

**DeJong, C. A.**
1998 Cutting tools & a high-speed chase. *Automotive Manufacturing & Production* 110(2):62-65.

**Deleersnyder, J., T. J. Hodgson, H. Muller and P. J. O'Grady**
1989 Kanban controlled systems: An analytic approach. *Management Science* 35(9):1079-91.

**Deming, W. E.**
1989 *Out of the Crisis.* Cambridge: MIT Center for Advanced Engineering Study.

**DeVor, R. E., T. H. Chang and J. Sutherland**
1992 *Statistical Quality Design and Control: Contemporary Concepts and Methods.* New York: Macmillan.

**Dudek, R. A., S. S. Panwalkar and M. L. Smith**
1992 The lessons of flowshop scheduling research. *Operations Research* 40(1):7-13.

**Duenyas, I.**
1999 Factory physics: The science of lean manufacturing. *Alumni Newsletter: Department of Industrial and Operations Engineering,* University of Michigan (Winter):6-7.

**Duenyas, I. and W. J. Hopp**
1995 Quoting customer lead times. *Management Science* 41(1):43-57.

**Ettlie, J. E. and J. D. Penner-Hahn**
1990 Focus, modernization and manufacturing technology policy. In J. E. Ettlie, M. C. Burnstein, and A. Fiegenbaum, eds., *Manufacturing Strategy*, pp. 153-64. Boston: Kluwer Academic Publishers.

**Ferdows, K.**
1997 Made in the world: The global spread of production. *Production and Operations Management* 6(2):102-9.

**Fleischer, M. and J. K. Liker**
1997 *Concurrent Engineering Effectiveness: Integrating Product Development Across Organizations.* Cincinnati: Hanser Gardner Publications.

**Flink, J. J.**
1981 Henry Ford and the triumph of the automobile. In C. W. Pursell, Jr., ed., *Technology in America: A History of Individuals and Ideas,* pp. 163-75. Cambridge: MIT Press.

**Flynn, B. B., S. Sakakibara and R. G. Schroeder**
1995 Relationship between JIT and TQM: Practices and performance. *Academy of Management Journal* 38(5):1325-60.

**Ford, H.**
1931 *Moving Forward.* London: William Heinemann Ltd.

**Ford, H. and S. Crowther**
1924 *My Life and Work.* London: William Heinemann Ltd..

**Frost, T. F.**
1994 Creating a teamwork-based culture within a manufacturing setting. *Industrial Management* 36(3):17-20.

**Galbraith, J. K.**
1958 *The Affluent Society.* Boston: Houghton Mifflin.

**Gallego, G. and I. Moon**
1992 The effect of externalizing setups in the economic lot scheduling problem. *Operations Research* 40(3):614-19.

**Gamow, G.**
1968 *Mr. Tompkins in Paperback.* Cambridge, England: Press Syndicate of the University of Cambridge.

**Grout, J. R.**
1997 Mistake-proofing production. *Production and Inventory Management Journal* 38(3):33-37.

**Gupta, Y. P. and S. C. Lonial**
1998 Exploring linkages between manufacturing strategy, business strategy, and organizational strategy. *Production and Operations Management* 7(3):243-64.

**Gupta, Y. P. and T. M. Somers**
1993 Manufacturing decisions and business strategy. *Manufacturing Review* 6(2): 87-100.

**Hammond, J. and J. Morrison**
1996 *The Stuff Americans Are Made Of.* New York: Macmillan.

**Harpell, J. L., M. S. Lane and A. H. Mansour**
1989 Operations research in practice: A longitudinal study. *Interfaces* 19(3):65-74.

**Harris, F. W.**
[1913] 1990 How many parts to make at once. *Factory: The Magazine of Management* 10(2):135-36, 152. Reprint, *Operations Research* 38(6):947-50.

**Hayes, R.**
1981 Why Japanese factories work. *Harvard Business Review* 59(4):57-66.

**Hayes, R. H. and G. P. Pisano**
1994 Beyond world-class: The new manufacturing strategy. *Harvard Business Review* 72(1):77-87.

**Hayes, R. and S. Wheelwright**
1984 *Restoring Our Competitive Edge: Competing Through Manufacturing.* New York: John Wiley & Sons, Inc.

**Hicks, P. E.**
1994 *Industrial Engineering and Management: A New Perspective.* New York: McGraw-Hill, Inc.

**Hinckley, C. M. and P. Barkan**
1995 The role of variation, mistakes, and complexity in producing nonconformities. *Journal of Quality Technology* 27(3):242-49.

**Hobbs, O. K.**
1994 Application of JIT techniques in a discrete batch job shop. *Production and Inventory Management Journal* 35(1):43-47.

**Hopp, W. J. and M. L. Spearman**
1996 *Factory Physics: Foundations of Manufacturing Management.* Chicago: Irwin.

**Hopp, W. J., M. L. Spearman and I. Duenyas**
1993 Economic production quotas for pull manufacturing systems. *IIE Transactions* 25(2):71-79.

**Hopp, W. J., M. L. Spearman and D. L. Woodruff**
1990 Practical strategies for lead time reduction. *Manufacturing Review* 3(2):78-84.

**Hopp, W. J., M. L. Spearman and R. Q. Zhang**
1997 Easily implementable inventory control policies. *Operations Research* 45(3): 327-40.

**Hounshell, D. A.**
1984 *From the American System to Mass Production, 1800-1932.* Baltimore: The Johns Hopkins University Press.

**Hurley, S. F. and S. Kadipasaoglu**
1999 Wandering bottlenecks: Speculating on the true causes. *Production and Inventory Management Journal* 39(4):1-4.

**Inman, R. A. and L. D. Brandon**
1992 An undesirable effect of JIT. *Production and Inventory Management Journal* 33(1):55-58.

**Inman, R. A. and S. Mehra**
1990 The transferability of just-in-time concepts to American small businesses. *Interfaces* 20(2):30-37.

**Inman, R. R.**
1993 Inventory is the flower of all evil. *Production and Inventory Management Journal* 34(4):41-45.

**Johnson, H. T. and R. S. Kaplan**
1987 *Relevance Lost: The Rise and Fall of Management Accounting.* Boston: Harvard Business School Press.

**Johnson, S. M.**
1954 Optimal two and three-stage production schedules with setup times included. *Naval Research Logistics Quarterly* 179(32):61-68.

**Jordan, W. C. and S. C. Graves**
1995 Principles on the benefits of manufacturing process flexibility. *Management Science* 41(4):577-94.

**Jusko, J.**
1998a The competitive edge. *IndustryWeek* 247(13):43-47.
1998b Beating the Joneses. *IndustryWeek* 247(22):27-33.
1998c Levels of dispute. *IndustryWeek* 247(22):35-38.

**Kane, V. E.**
1986 Process capability indices. *Journal of Quality Technology* 18(1):41-52.

**Kannan, V. R.**
1998 Analysing the trade-off between efficiency and flexibility in cellular manufacturing. *Production Planning & Control* 9(6):572-79.

**Kim, G. C. and E. Takeda**
1996 The JIT philosophy is the culture in Japan. *Production and Inventory Management Journal* 37(1):47-50.

**Kimura, O. and H. Terada**
1981 Design and analysis of pull systems, a method of multi-stage production control. *International Journal of Production Research* 19(3):241-53.

**Kingman, J. F. C.**
1961 The single server queue in heavy traffic. *Proceedings of the Cambridge Philosophical Society* 57:902-4.

**Kinni, T. B.**
1996 *America's Best: IndustryWeek's Guide to World-Class Manufacturing Plants.* New York: John Wiley & Sons, Inc.

**Kirkpatrick, D.**
1997 Now everyone in PCs wants to be like Mike. *Fortune* 136(5):91.

**Kotter, J. P.**
1996 *Leading Change.* Boston: Harvard Business School Press.

**Koufteros, X. A. and M. A. Vonderembse**
1998 The impact of organizational structure on the level of JIT attainment: Towards theory development. *International Journal of Production Research* 36(10):2863-78.

**Krafcik, J. F.**
1988 Triumph of the lean production system. *Sloan Management Review* 30(1):41-52.

**Kuik, R. and P. F. J. Tielemans**
1997 Setup utilization as a performance indicator in production planning and control. *International Journal of Production Economics* 49(2):175-82.

**Kumon, H.**
1994 International transferability of the Japanese production system: Japanese-affiliated auto plants in the U.S.A., the U.K. and Taiwan. *Journal of International Economic Studies* 8:59-78.

**Lampel, J. and H. Mintzberg**
1996 Customizing customization. *Sloan Management Review* 38(1):21-30.

**Lancaster, H.**
1999 There's a new boss at the top; where does that leave you? *The Wall Street Journal,* 19 January.

**Layden, J.**
1998 The reality of APS systems. *APICS—The Performance Advantage* 8(9):50-52.

**Lee, H. L., V. Padmanabhan, and S. Whang**
1997a The bullwhip effect in supply chains. *Sloan Management Review* 38(3):93-102.
1997b Information distortion in a supply chain: The bullwhip effect. *Management Science* 43(4):546-58.

**Lee, H. L. and M. J. Rosenblatt**
1987 Simultaneous determination of production cycle and inspection schedules in a production system. *Management Science* 33(September):1125-36.

**Leong, G. K. and P. T. Ward**
1995 The six Ps of manufacturing strategy. *International Journal of Operations and Production Management* 15(12):32-45.

**Leschke, J. P.**
1998 A new paradigm for teaching introductory production/operations management. *Production and Operations Management* 7(2):146-59.

**Li, G. and S. Rajagopalan**
1998 Process improvement, quality, and learning effects. *Management Science* 44(11):1517-32.

**Liker, J. K., ed.**
1997 *Becoming Lean: Inside Stories of U.S. Manufacturers.* Portland, OR: Productivity Press.

**Little, J. D. C.**
1961 A proof of queuing formula: L=λW. *Operations Research* 9(3):383-87.
1992 Tautologies, models and theories: Can we find "laws" of manufacturing? *IIE Transactions* 24(3):7-13.

**Love, C. E., R. Guo and K. H. Irwin**
1995 Acceptable quality level versus zero-defects: Some empirical evidence. *Computers & Operations Research* 22(4):403-17.

**Maital, S.**
1991 The profits of infinite variety. *Across The Board* 28(10):7-10.

**Marine, A., and P. Riley**
1995 Creating a culture of change. *Hospital Materiel Management Quarterly* 16(4): 30-40.

**Martin, J.**
1997 Give 'em exactly what they want. *Fortune* 136(9):283.

**McWilliams, G.**
1999 Dell Computer takes three winning spots in an unrivaled feat. *The Wall Street Journal,* 25 February.

**Medhi, J.**
1991 *Stochastic Models in Queueing Theory.* Boston: Academic Press.

**Michael, J.**
1997 A conceptual framework for aligning managerial behaviors with cultural work values. *International Journal of Commerce & Management* 7(3/4):81-101.

**Montgomery, D. C.**
1991 *Introduction to Statistical Quality Control.* 2nd. ed. New York: John Wiley & Sons, Inc.

**Montgomery, D. C. and C. M. Mastrangelo**
1991 Some statistical process control methods for autocorrelated data. *Journal of Quality Technology* 23(3):179-93.

**Morgan, G.**
1986 *Images of Organization.* Newbury Park, CA: Sage Publications.

**Moskal, B. S.**
1995 General Motors. *IndustryWeek* 244(19):33-34.

**Muckstadt, J. A. and S. R. Tayur**
1995 A comparison of alternative kanban control mechanisms. I. Background and structural results. *IIE Transactions* 27(2):140-50.

**Murphy, T.**
1998 Close enough to perfect. *Ward's Auto World* 34(8):50.

**Myerson, A. R.**
1996 O Governor, won't you buy me a Mercedes plant? *The New York Times*, 1 September.

**Nahmias, S.**
1997 *Production and Operations Analysis.* 3$^{rd}$ ed. Chicago: Irwin.

**Nelson, E.**
1994 Truck sales shift into high. *Business Marketing* 79(4):3, 41.

**New York Times, The**
1999 The view from the outside: Levi's needs more than a patch. 28 February.

**Nicholas, J. M.**
1998 *Competitive Manufacturing Management.* Boston: Irwin/McGraw-Hill.

**Ocana, C. and E. Zemel**
1996 Learning from mistakes: A note on just-in-time systems. *Operations Research* 44(1):206-14.

**Orlicky, J.**
1975 *Material Requirements Planning: The New Way of Life in Production and Inventory Management.* New York: McGraw-Hill.

**Picken, J. C. and G. G. Dess**
1996 The seven traps of strategic planning. *Inc.* 18(11):99.

**Pilkington, A.**
1998 Manufacturing strategy regained: Evidence for the demise of best-practice. *California Management Review* 41(1):31-42.

**Pollack, A.**
1999 Aerospace gets Japan's message. *The New York Times*, 9 March, late edition.

**Porteus, E. L.**
1986 Optimal lot sizing, process quality improvement, and setup cost reduction. *Operations Research* 34(1):137-44.

**Pursell, C. W., Jr.**
1981a Technology in America: An introduction. In C. W. Pursell, Jr., ed., *Technology in America: A History of Individuals and Ideas,* pp. 1-7. Cambridge: MIT Press.
1981b Cyrus McCormick and the mechanization of agriculture. In C. W. Pursell, Jr., ed., *Technology in America: A History of Individuals and Ideas,* pp. 71-79. Cambridge: MIT Press.

**Ramstad, E.**
1998 Compaq stumbles as PCs weather new blow. *The Wall Street Journal,* 9 March.

277

**Rishel, T. D. and O. M. Burns**
1997 The impact of technology on small manufacturing firms. *Journal of Small Business Management* 35(1):2-10.

**Robinson, A.**
1991 Origins of the modern Japanese management style. In A. Robinson, ed., *Continuous Improvement in Operations: A Systematic Approach to Waste Reduction,* pp. 1-26. Cambridge: Productivity Press.

**Robson, R. E.**
1994 The Shingo Prize – what it is, and who won this year. *Tapping the Network Journal* 5(3):23-28.

**Rother, M. and J. Shook**
1998 *Learning to See.* Brookline, MA: The Lean Enterprise Institute.

**Sakakibara, S., B. B. Flynn and R. G. Schroeder**
1993 A framework and measurement instrument for just-in-time manufacturing. *Production and Operations Management* 2(3):177-94.

**Sakakibara, S., B. B. Flynn, R. G. Schroeder and W. T. Morris**
1997 The impact of just-in-time manufacturing and its infrastructure on manufacturing performance. *Management Science* 43(9):1246-57.

**Sandelands, E.**
1995 Great expectations for lean suppliers. *International Journal of Physical Distribution & Logistics* 24(3):40-42.

**Schein, E. H.**
1986 What you need to know about organizational culture. *Training and Development Journal* 40(1):30-33.

**Schonberger, R. J.**
1986 *World Class Manufacturing: The Lessons of Simplicity Applied.* New York: The Free Press.
1990 *Building a Chain of Customers: Linking Business Functions to Create a World Class Company.* New York: The Free Press.
1996 Backing off from the bottom line. *Executive Excellence* 13(5):16-17.

**Schroeder, R. G., J. C. Anderson and G. Cleveland**
1986 The content of manufacturing strategy: An empirical study. *Journal of Operations Management* 6(4):405-15.

**Serwer, A. E.**
1997 Michael Dell turns the PC world inside out. *Fortune* 136(5):76-86.

**Shapiro, E. C.**
1995 *Fad Surfing in the Boardroom: Reclaiming the Courage to Manage in the Age of Instant Answers.* Reading, MA: Addison-Wesley Publishing Company.

**Sharf, S.**
1999 A sardine story: There's more to it than just buying and selling. *Ward's Auto World* 35(2):21.

**Shenkman, R.**
1994 *Legends, Lies, and Cherished Myths of American History.* Reprint ed. New York: Harper Perennial.

**Sheridan, J. H.**
1996 The global economic engine. *IndustryWeek* 245(10):16-24.

**Shingo, S.**
1989 *A Study of the Toyota Production System from an Industrial Engineering Viewpoint.* Cambridge: Productivity Press.

**Shook, J. Y.**
1997 Bringing the Toyota Production System to the United States: A personal perspective. In J. K. Liker, ed., *Becoming Lean: Inside Stories of U.S. Manufacturers,* pp. 40-69. Portland, OR: Productivity Press.

**Simison, R. L.**
1998 Fears of overcapacity continue to grow—auto glut worries analysts despite some cutbacks due to turmoil in Asia. *The Wall Street Journal,* 2 March.

**Skinner, W.**
1969 Manufacturing—missing link in corporate strategy. *Harvard Business Review* 47(3):136-45.

**Smith, M.**
1998 Culture and organisational change. *Management Accounting-London* 76(7):60-62.

**South Carolina Department of Commerce**
1998 Division of Research and Grants. *Business Incentives,* Columbia.

**Spearman, M. L.**
1997 On the theory of constraints and the goal system. *Production and Operations Management* 6(1):28-33.

**Spearman, M. L. and M. A. Zazanis**
1992 Push and pull production systems: Issues and comparisons. *Operations Research* 40(3):521-32.

**Spira, J. S. and B. J. Pine**
1993 Mass customization. *Chief Executive* 83:26-29.

**Spoor, L.**
1998 Metrics: Process improvement through measurement. *APICS-The Performance Advantage* 8(11):54-56.

**Standard, C. T.**
1997 Process control in the presence of autocorrelation. In B. Bidanda and S. Jagdale, eds., *6th Industrial Engineering Research Conference Proceedings,* pp. 101-6. Norcross, GA: Institute of Industrial Engineers.

**Standard, C. T. and D. Davis**
1999 Lean thinking: is thinking required? *Automotive Engineering International* 107(7).

**Stewart, T. A.**
1995 Review of Fad surfing in the boardroom, by E. C. Shapiro. *Fortune* 132(10):162.

**Struebing, L.**
1997 Kaizen pays off for manufacturers. *Quality Progress* 30(4):16.

**Sugimori, Y., K. Kusunoki, F. Cho and S. Uchikawa**
1977 Toyota Production System and kanban system—materialization of just-in-time and respect-for-human system. *International Journal of Production Research* 15(6):553-64.

**Suri, R., G. W. W. Diehl, S. de Treville and M. J. Tomsicek**
1995 From CAN-Q to MPX: Evolution of queuing software for manufacturing. *Interfaces* 25(5):128-50.

**Suzaki, K.**
1987 *The New Manufacturing Challenge.* New York: The Free Press.

**Tannenbaum, J. A.**
1997 Mexican-food joint venture gives Arby's indigestion. *The Wall Street Journal,* 12 August.

**Taylor, A.**
1996 If you cut waste, you win. *Fortune* 134(12):213-14.

**Temponi, C. and S. Y. Pandya**
1995 Implementation of two JIT elements in small-sized manufacturing firms. *Production and Inventory Management Journal* 36(3):23-29.

**Tucker, M. W. and D. A. Davis**
1993 Key ingredients for successful implementation of just-in-time: A system for all business sizes. *Business Horizons* 36(3):59-65.

**Tunis, E.**
1965 *Colonial Craftsmen and the Beginnings of American Industry.* Cleveland: The World Publishing Company.

**U.S. Bureau of the Census**
1996 *Census of Manufacturers,* 1992. Bureau of the Census. Washington, D.C.

**Wallechinsky, D. and I. Wallace**
1975 *The People's Almanac.* Garden City, NY: Doubleday & Company, Inc.

**War Manpower Commission**
1945 *The Training Within Industry Report.* Washington, D.C.: GPO.

**Ward, P. T., D. J. Bickford and G. K. Leong**
1996 Configurations of manufacturing strategy, business strategy, environment and structure. *Journal of Management* 22(4):597-626.

**Warnecke, H. J. and M. Huser**
1995 Lean production. *International Journal of Production Economics* 41(1-3):37-43.

**Whiteside, D. and J. Arbose**
1984 Unsnarling industrial production: Why top management is starting to care. *International Management* 39(3):20-26.

**Wilson, J. M.**
1996 Henry Ford: A just-in-time pioneer. *Production and Inventory Management Journal* 37(2):26-31.
1998 A comparison of the 'American System of Manufactures' circa 1850 with just in time methods. *Journal of Operations Management* 16(1):77-90.

**Womack, J. P.**
1993 Mass customization: The new frontier in business competition. *Sloan Management Review* 34(3):121-22.

**Womack, J. P. and D. T. Jones**
1996 *Lean Thinking*. New York: Simon & Schuster.

**Womack, J. P., D. T. Jones and D. Roos**
1990 *The Machine That Changed the World*. New York: Harper Perennial.

**Woolsey, G.**
1993 Where were we, where are we, where are we going, and who cares?" *Interfaces* 23(5):40-46.

**Wrennall, W. and M. Markey**
1995 Kanban material acquisition system. *Management Services* 39(1):18-21.

**Wright, K.**
1998 The shape of things to go. *Scientific American* 262(5):92-99.

**Yamada, Y.**
1998 Conversation. Indianapolis, IN, March.

**Yanagawa, Y., S. Miyazaki and H. Ohta**
1994 An optimal operation planning for the fixed quantity withdrawal kanban system with variable leadtimes. *International Journal of Production Economics* 33(1): 163-68.

**Zipkin, P. H.**
1995 Performance analysis of a multi-item production-inventory system under alternative policies. *Management Science* 41(4):690-703.

# –A–

ABC (see Activity Based Costing)
activity based costing, 232
Adams, William, 2
advanced planning and scheduling, 164
American Production and Inventory Control Society, 67, 162
American Supplier Institute, 67
American System, 6–7, 22, 49–50, 58–59, 68
American System of Manufacturing, 13, 15
APICS (see American Production and Inventory Control Society)
APS (see Advanced Planning and Scheduling)
Arkwright, Richard, 13
ARL (see Average Run Length)
artisan workshops, 11
artisans, 11
ASI (see American Supplier Institute)
assembly line production, 17
authority, 65
autocorrelation, 191
availability, 85–86
average run length, 192

# –B–

B. F. Goodrich, 57
batch and push production (see production)
batch and queue production (see production)
batch production (see production)
batch size, 110
best case performance (see performance)
best-practice strategy (see strategy)
Boeing, 56–57
bottleneck, 112
bottleneck analysis, 42
bottleneck wandering, 35
Brown, Moses, 14
buffers, 23
built-in quality (see quality)
bullwhip effect, 141, 144
business goals, 97

# –C–

# –D–

Day, Joseph, 54
defects, 133, 217
Dell Computer Corporation, 31

# –E–

Economic Order Quantity, 210–12
economies of scale, 16, 51, 66, 71
effective processing time, 86
effective *scv* (see squared coefficient of variation)
employee responsibility, 65
ending, the, 246
enterprise resource planning, 66, 163
EOQ (see Economic Order Quantity)
ERP (see Enterprise Resource Planning)
Evans, Oliver, 17

# –F–

factory culture (see culture)
factory physics, 47, 74, 84–85, 87–88, 117, 234
factory physics, laws of, 92
factory transformation (see transformation)
false alarm, 192
financial measures, 255
fitness for use, 126
Flanders, Walter E., 16
flexibility, 57, 138, 218
Ford, Henry, 16, 58
Ford, Highland Park, 17, 58, 68
Ford Model T, 17–19, 58
Ford Motor Company, 58
Ford Rouge complex, 60
formal organization structure, 264
Freudenburg-NOK, 54, 56
functional gauging, 126
future state, 245

# –G–

GD&T (see Geometric Dimensioning and Tolerancing)
geometric dimensioning and tolerancing, 127
goals, corporate, 95–97
guiding coalition, 263

# –H–

*heijunka* (see level, mixed production)
heirarchical culture, 259
heirarchies and clans, 259
Highland Park (see Ford)
human resources, 23

# –I–

improvement cycle, 128
incidental work, 62, 65
industrial revolution, 13
informal organization structure, 264
in-line sequencing, 140
in-line vehicle sequencing, 140
Inman, Robert, 100
innovation, 264
inspection, 185, 203
inspection, informative, 185–86
inspection, judgement, 185–186
inspection, self-, 187
inspection, successive, 187
interchangeable parts, 14
inventory, 23–24, 107, 109, 121, 132, 169
inventory, excess, 217
inventory reduction, 137

# –J–

Japan, 59, 69–71
Japanese manufacturing techniques, 68
*jidoka* (see quality at-the-source)
JIT (see just-in-time)

JIT production (see just-in-time production)
job rotation, 264
job shop scheduling, 43
Johnson & Johnson, 55
just-in-time, 2, 24, 34, 58, 68, 70, 94, 123, 137–38, 213
just-in-time delivery, 155, 157–58
just-in-time production, 147, 157–59, 161, 251

## –K–

*kaizen*, 119, 134
*kaizen* event, 56, 127
*kaizen*, paintball, 135
*kanban*, 170, 173, 177
Krafcik, John, 49

## –L–

Lacandon Maya, 258
large lot production (see production)
laws of manufacturing, 6
lead time reduction, 97, 114
lean, 49
lean concepts, 58
lean manufacturing, 3–4, 6–7, 47, 50, 52, 53, 54, 57, 59, 65, 67–69, 71, 99, 234
lean manufacturing. philosophy, 59, 70–71, 92, 117
lean manufacturing strategy (see strategy)
lean manufacturing techniques, 223
lean practices, 57, 207
lean principles, 72, 100
lean production (see production)
lean production values, 70
lean techniques, 3
lean thinking, 1, 7, 73
lean transformation (see transformation)
learning organizations, 110
level, mixed production, 138
level, sequential flow, 140
Liker, Jeffrey, 54

line balancing, 152–55
Little, John D., 78
Little's Law, 78, 80, 106, 110
lot size, 217–18, 222
lot-size reduction, 251
Lowell, Francis Cabot, 14

# –M–

maintenance, 229
maintenance department, 85–87
Malcolm Baldridge Award, 179
management commitment, 254–55
management policy, 219
management style, 263
manufacturing cell, 111, 264
manufacturing, cellular, 23, 78, 79, 150
manufacturing challenge, 11, 27
manufacturing, science of, 1, 6, 47, 75
manufacturing strategy (see strategy)
manufacturing system, 117
marketplace, 171
mass customization, 21
mass production (see production)
mass production values, 70–71
master craftsman, 12
material handling, 156
material requirements planning, 34, 66, 161–62
mechanized conveyance, 17
milk run, 157
misconceptions, 67
mistake-proofing, 181, 195, 197, 202
mistake-proofing devices, 199
MIT International Motor Vehicle Program, 49
mixed-model production, 140
Model T (see Ford)
motivation, 132, 183, 219, 248
MRP (see Material Requirements Planning)
MRP-II, 66
multiskilled workers, 57

# –N–

neutral zone, 245, 248
new beginning, 245, 248
Nissan, 10
non-polynomial-hard, 46
non-value-added activity, 62
Northrop Grumman, 56
NP-hard (see non-polynomial-hard), 46
NUMMI, 49

# –O–

Ohno, Taiichi, 59–60, 100
one-piece flow production (see production)
operational stability, 117–118, 124
organizational structure, 54, 263
overhead absorption, 230
overproduction (see production)

# –P–

pace of production (see production)
PCR (see Process Capability Ratio)
perceptual templates, 52
performance, best case, 78, 80
performance index, 238
performance measures, 32, 225, 227–29, 239, 253
performance, worst case, 77–78, 80
philosophy of manufacturing, 6, 50, 61
philosophy of production (see production)
pig-in-a-python effect, 111
process capability ratio, 193
process improvement, 127
process-oriented layout, 23
process villages, 23
processing time, 4, 215, 219
producer's risk, 191
product family, 244
product oriented, 22
production, batch, 209

## –Q–

## –R–

# –V–

value-added activity, 62, 64–65
value stream, 4, 73, 88, 90–91, 102
value-stream mapping, 135, 173
value system, 51, 70–71
values, 51
variability, 4–5, 23–24, 43, 82–83, 87, 90–92, 109, 123, 134, 165, 219–20, 223, 235, 236, 238
variability in cycle time, 111
variability reduction, 5, 112, 130, 138
vertical integration, 6, 14

# –W–

waiting time, 62, 72
waste, 75–76, 130, 134
wasted effort, 65
Whitney, Eli, 14
Whiz Kids, 32
WIP (see Work In Process)
WIP cap, 164
Wollering, Max F., 16
work cells, 58, 84–85, 152, 154, 247
work in process, 53, 76, 78–80, 101
work in process, constant, 167
work standardization, 57
workload, 112

# –Z–

zero defects, 181, 189, 199, 203
zero inventory, 159
zero inventory policy, 137
zero inventory systems, 164